The Little Book of Math into English

George Grätzer

The Little Book
of Math into English

 Springer

George Grätzer
Toronto, ON, Canada

ISBN 978-3-031-65160-1   ISBN 978-3-031-65161-8 (eBook)
https://doi.org/10.1007/978-3-031-65161-8

This Springer imprint is published by the registered company Springer Nature Switzerland AG
The registered company address is: Gewerbestrasse 11, 6330 Cham, Switzerland

If disposing of this product, please recycle the paper.

Contents

Introduction

Stepping away from the daunting lengths of typical English grammar books, which span from 400 to 1800 pages,[1] this guide presents a concise and mathematician-friendly approach. My goal isn't to overwhelm you with extensive grammar rules, but provide key insights that are directly relevant and easily applicable.

Designed specifically for mathematicians (or physicists), I avoid the complex jargon of standard grammar guides. I aim to simplify the process by selecting a handful of essential topics, ranging from direct and indirect sentences to INTROS—a new concept, key to writing readable math. By following our straightforward recommendations, you can reduce up to 80% of common errors in your papers,

Figure 1: Fat grammar books (Vecteezy Library)

thereby enhancing their clarity and readability. Don't worry about achieving 100% perfection—that would require a 600-page manual!

I added 'extras' (which you can reach by a link to this book's Springer Web page) to each topic, with the title: *Practice makes perfect*. Do some exercises to see what you have learned from the topic. The *Hints* give you a helping hand.

I suggest using a manageable strategy, reading one topic a day. It's just a few minutes of easy reading that can significantly improve your writing skills. Some 400 examples help you along; grammatically correct examples are marked by a 'smiling meme' 😊 and the others by a 'sad meme' 😞.

---

[1]For instance, see item [14] in the Bibliography.

"What is the use of a book without pictures or conversations?" Alice thought in Lewis Carroll's work [3]. To lighten the reading, I incorporated many illustrations and also some conversations (with the generative artificial intelligence system ChatGPT).

I asked ChatGPT:

I am writing a book to improve the use of English in math articles. Am I crazy?
And the answer:

Not at all! Writing a book to teach better English for math expositions is an excellent idea and addresses a vital need.

Encouraging indeed.

For whom the bell tolls

(This is the title of Ernest Hemingway's famous novel [12]. The quote is from John Donne's Meditation XVII.) This book is primarily intended for mathematicians whose native language is not English, though native English speakers will find value in some sections. Where

Figure 2: Illustration library (Vecteezy Library)

American and British customs diverge, I adhere to American conventions to maintain simplicity.

The sum of my parts

Throughout my career as a mathematician, I have published approximately 300 articles and authored 34 books; see [9]. My scholarly contributions have earned over 3,425 citations. In 1971, I founded the journal Algebra Universalis, which has since published 85 volumes and now comprises around 34,000 pages. This extensive body of work has provided me with numerous opportunities to observe both the misuse and proper use of English by mathematicians.

The illustrations in this book are meant to provide *visual relief* from pages and pages of text. Apart from a few photos, most are from the Vecteezy illustration library, housing millions. Subscribers have unlimited downloads and full commercial rights with no required attribution.

This *Introduction* is followed by *Overview*, listing the 23 topics covered in the book, with a brief description for each. I strongly recommend reading this to get an overview of the book, so if you need help, you know where to turn.

In Appendix C, the essay, "How to write mathematics" by P. R. Halmos—a master expositor—deals more with the math aspects of writing; see [11]. It is reprinted with permission as an addendum.

Have fun and enjoy this journey toward clearer, more effective math writing.

An extremely talented and experienced group of experts—Barbara Beeton, Gregory L. Cherlin, Gábor Czédli, David Derbes, Michael Doob, and Murray Eisenberg—read the draft and provided detailed reports. I am indebted to them.

George Grätzer

On the Ides of March, 2024, in Toronto

Overview

Topic 1. Little words I: Prepositions

This chapter discusses the correct usage of prepositions in math contexts. It explains how prepositions like 'between', 'among', and 'across' function to clarify relationships and positions within math arguments, and provides tips on avoiding common prepositional mistakes that can lead to ambiguity.

Topic 2. Little words II: Pronouns

Focusing on pronouns, this chapter emphasizes the need for clarity and precision to avoid confusion in math texts. It addresses the potential ambiguities introduced by pronouns and offers strategies for ensuring that references are clear and unambiguous.

Topic 3. Little words III: Conjunctions

The use of conjunctions in linking math statements is explored here. The chapter illustrates how proper use of conjunctions like 'and', 'but', and 'or' can enhance the logical flow of arguments and help maintain clarity in complex math discussions.

Topic 4. Little words IV: So, such, that

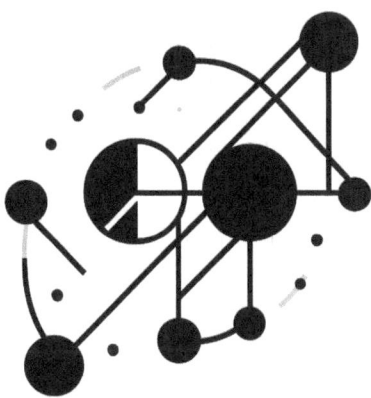

Figure 3: Conjunctions (Vecteezy Library)

This chapter delves into the nuanced use of 'so', 'such that', and 'that' to express consequence, purpose, and specification in math writing. It advises on avoiding overuse and ensuring these words contribute to clarity and precision in stating math results or setting up equations and proofs.

Topic 5. Little words V: Like, such as

Here, the distinction between 'like' for comparisons and 'such as' for specifying examples is clarified. The chapter provides guidance on using these terms to enhance specificity and accuracy in math descriptions, with examples to illustrate correct usage.

Topic 6. Little words VI: Either, or, both

Discusses the proper use of 'either', 'or', and 'both' in expressing alternatives and conditions in math texts. It includes examples that demonstrate how to use these words to clearly convey exclusivity or co-occurrence of conditions in math reasoning.

Topic 7. Little marks I: Punctuation

This chapter covers the essential role of punctuation in math writing, from commas and semicolons to colons and quotation marks. It offers rules and examples for using punctuation to enhance readability and clarity, emphasizing the importance of the Oxford comma.

Topic 8. Little marks II: Hyphens

Focused on hyphens, this chapter provides rules for their use in math writing, particularly in compound terms and when used with prefixes and suffixes. It explains how hyphens can prevent ambiguity and ensure clear communication in math expressions.

Topic 9. Little marks III: Dashes

Explains the differences between en-dashes and em-dashes and their appropriate uses in math writing. This chapter shows how dashes can be used to denote ranges, connect items, or add emphasis within math discussions.

Topic 10. Little transitions: Short version

This chapter focuses on the use of transitional words to connect ideas smoothly in math writing. It introduces basic transitional phrases that help in building logical sequences and clarifying relationships among different math concepts.

Topic 11. Little transitions: Longer version

Offers a comprehensive guide to using transitional phrases in math writing. This chapter provides deeper insights into how effectively used transitions can guide the reader through complex arguments and enhance the overall coherence of math texts.

Topic 12. Little INTROs

The concept of INTROs, or introductory phrases used before introducing formulas or key terms, is discussed. This chapter emphasizes how INTROs can aid in readability and comprehension, providing practical examples of their effective use in math writing.

Topic 13. Little Run: Run-on Sentences

This chapter addresses the issue of run-on sentences in math writing, explaining their impact on clarity and readability. It provides strategies for identifying and correcting run-on sentences, including the use of proper punctuation and conjunctions to separate independent clauses. The emphasis is on creating concise and understandable sentences to improve the logical flow of math discourse.

Topic 14. Little Trip: To go or going

Discusses the correct use of the infinitive and gerund forms of verbs in math writing, explaining how these forms affect the clarity and tone of the text. Examples illustrate when to use each form to express actions or ongoing processes effectively. The chapter helps writers choose the appropriate verb form to maintain precision in instructions and descriptions.

Topic 15. Little dangling

This chapter deals with dangling modifiers in math writing, offering revision strategies to align modifiers clearly with the words they are intended to describe. Examples show how misplaced modifiers can confuse readers and distort the intended meaning of sentences.

The chapter provides practical tips for rephrasing sentences to ensure clarity and accuracy.

Topic 16. Little numbers: Less and fewer, both and two

Clarifies the correct usage of the quantifiers 'less' and 'fewer', and explains the distinction between 'both' and 'two' in math contexts. The chapter addresses common grammatical errors and provides rules to help writers choose the right word based on context and meaning. It aims to prevent numerical and logical inaccuracies in math writing.

Topic 17. Little Pause: Comma before if

Provides specific guidance on using commas before conditional 'if' clauses in math texts. This chapter explains how such punctuation affects clarity and meaning, with examples showing correct and incorrect comma usage. The nuances of introducing conditional statements are explored to help writers avoid common punctuation mistakes.

Topic 18. Little ambiguity

Focuses on recognizing and resolving ambiguous statements in math writing to prevent misinterpretation and errors in logical reasoning. The chapter provides strategies for clear and precise expression, with examples of ambiguous scenarios and their clarifications. It encourages writers to scrutinize their work for potential ambiguities and refine their wording accordingly.

Topic 19. Little active vs. passive

Discusses the impact of voice on the clarity and directness of math writing, recommending the active voice for enhanced engagement and clarity. The chapter contrasts examples of active and passive constructions, demonstrating how active voice can make explanations more direct and easier to follow.

Figure 4: Run-on sentence (Vecteezy Library)

Topic 20. Little speed: Faster and fastests

Offers tips on writing concisely and clearly to communicate math ideas efficiently, avoiding common pitfalls in verbose explanations. The chapter encourages
the use of straightforward language and the elimination of unnecessary details that
can slow down the reader's understanding.

Topic 21. Little distinctions

Delves into the subtle differences between commonly confused terms in math
language. This chapter is structured to enhance precision and professional expression
by addressing these distinctions:

Section 21.1: 'a' vs. 'the' vs. blank

Explores the nuanced use of articles in math writing, explaining when to choose
'a', 'the', or no article at all to ensure clarity and specificity in descriptions and
arguments.

Section 21.2: Basics

Covers fundamental distinctions between terms that often get mixed up, providing a foundation for understanding the proper usage of critical vocabulary in math
contexts.

Section 21.3: Typical mistakes

Highlights common errors made due to misunderstanding or misapplying terms,
offering corrective advice to avoid these pitfalls in professional math writing.

Section 21.4: More examples

Presents additional examples to illustrate correct versus incorrect uses of terms,
reinforcing the lessons from earlier sections through practical application.

Section 21.5: Maybe vs. may be

Clarifies the difference between 'maybe' (an adverb meaning perhaps) and 'may
be' (a verb phrase indicating possibility), which are often confused in writing.

Section 21.6: Further vs. farther

Distinguishes between 'further' (abstract or figurative distance) and 'farther'
(physical distance), with guidelines on their appropriate contexts.

Section 21.7: Shall vs. will

Explains the traditional and modern uses of 'shall' and 'will', guiding the proper

Figure 5: Foreword vs. forward (Vecteezy Library)

choice based on formality and intended meaning.

Section 21.8: Foreword vs. forward

Differentiates 'foreword' (a prefatory section of a book) from 'forward' (directional or progressive movement), which are often misused.

Section 21.9: Not vs. rather than

Discusses the preference for 'not' over 'rather than' in certain contexts to avoid ambiguity and improve the flow of math reasoning.

Section 21.10: Cannot vs. can not

Details the subtle difference in usage between 'cannot' (denoting impossibility) and 'can not' (optional negation), which can significantly impact the clarity of statements.

Section 21.11: Note vs. notice

Clarifies when to use 'note' (to observe or mention something important) versus 'notice' (to become aware of), which can enhance the precision of communication.

Section 21.12: Effect vs. affect

Addresses the common confusion between 'effect' (a noun meaning result) and 'affect' (a verb meaning to influence), a crucial distinction for accurate scientific writing.

Topic 22. Little Modification

Delves into modifying nouns and verbs in math language to refine and specify meanings more precisely. This chapter offers insights into the subtle nuances of word choice and modification that can significantly impact the clarity and accuracy of math statements.

Topic 23. Little Only

Focuses on the placement and impact of the word 'only' in math sentences, highlighting how its position can significantly change the meaning of a statement. The chapter provides examples to show the correct positioning of 'only' to convey the intended meaning accurately, helping writers avoid common misplacements that can alter the interpretation of math claims.

Figure 6: Only (Vecteezy Library)

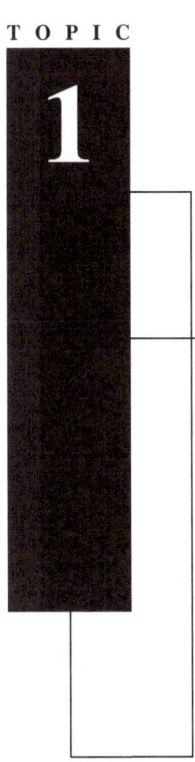

Little words I. Prepositions

Prepositions are small yet powerful words that link nouns, pronouns, or phrases to other words within a sentence. Prepositions help convey relationships between different entities or ideas, such as dimensions, sets, functions, and logical constructs.

1.1 Common prepositions in math

In indicates membership within sets or domains.

> *Consider a point in a plane.*

Between expresses a relationship involving two entities with something in the middle.

> *The function achieves its maximum between these two points.*

Among is used when the relationship involves more than two entities.

> *Among the many proofs of the Pythagorean Theorem, one stands out.*

Across indicates a span or extension over something.

> *We apply the transformation across the entire dataset.*

Supplementary Information The online version contains supplementary material available at https://doi.org/10.1007/978-3-031-65161-8_1.

Figure 1.1: Prepositions: in and under (Vecteezy Library)

Beyond is used to describe a location on the farther side of a specified boundary.

> *The sequence extends beyond any finite bound.*

With respect to is used to describe a relationship of comparison or differentiation, commonly found in calculus and differential equations.

> *We want to find the derivative of the function $f(x) = x^2$ with respect to x.*

Under is used to indicate a condition or premise.

> *Under the assumption of continuity, the function behaves predictably.*

1.2   *Using prepositions correctly*

Be careful to use them correctly.

> *The limit approaches to infinity.*
>
> *The limit approaches infinity.*
> The correct expression does not require 'to' after 'approaches'.

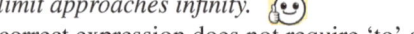

The derivative is calculated on $x = 3$.

The derivative is calculated at $x = 3$.

When referring to a specific point, 'at' is the correct preposition.

————

The function converges in zero.

The function converges to zero.

Use 'converges to' when discussing the behavior of a function approaching a particular value.

————

We integrate the function by 0 to 1.

We integrate the function from 0 to 1.

When specifying the interval of integration, 'from... to...' is the correct structure.

————

The series expansion is valid onto a certain radius of convergence.

The series expansion is valid within a certain radius of convergence.

'Within' describes being inside the limits of the radius of convergence.

————

The probability of the event is depending on several variables.

The probability of the event depends on several variables.

'Depends on' is the correct phrase to indicate dependency, not 'depending on'.

————

The matrix is orthogonal for the given basis.

The matrix is orthogonal with respect to the given basis.

'With respect to' is the proper phrase when discussing how a property relates to something else.

————

The graph of the function intersects in the y-axis.

The graph of the function intersects the y-axis.

'Intersects' doesn't need a preposition when stating what it intersects with.

————

The eigenvalues are belonging under a specific operator.

The eigenvalues belong to a specific operator.

'Belong to' is the correct phrase to indicate possession or association.

————

The solution to the differential equation exists inside a specific domain.

The solution to the differential equation exists over a specific domain.

'Over' is often used to discuss the domain where a solution is applicable, though 'in' can also be correct depending on context.

1.3 *Placement*

Traditional guidelines suggest that a preposition should never be used to end a sentence.[1] However, in practice, ending a sentence with a preposition can make the sentence more natural and avoid awkward constructions. For example:

The function with which we are dealing... 🙁

The function we are dealing with... 🙂

1.4 *Prepositions with math terms*

On is often used when referring to functions acting on a set or when discussing surfaces or boundaries.

Plot the points on the Cartesian plane. 🙂

The line on the graph represents the function's slope. 🙂

With can describe possession or characteristics, as in equations or properties associated with math objects.

Solve the equation with variables on both sides. 🙂

———

She came up with a solution with the help of calculus. 🙂

By is used to indicate division or differentiation, especially in math operations.

Divide 10 by 2 to get 5. 🙂

———

The area of a rectangle is found by multiplying its length by its width. 🙂

1.5 *Avoiding ambiguity*

Ambiguity in math writing, especially due to the placement of prepositions, can be problematic.

Consider the elements added to set A from set B. 🙁

Consider adding all elements of set B to set A. 🙂

The ambiguity arises from whether elements are being selected from set B to be added to A, or if all of set B is to be combined with A. In formula, $\{\, a + b \mid a \in A, b \in B \,\}$, on the one hand, and $\{\, A + b \mid b \in B \,\}$, on the other.

———

The function is increasing for values greater than zero. 🙁

The function is increasing on the set of positive elements in its domain. 🙂

---

[1]In March 7, 2024, linguist John McWhorter wrote in the New York Times: "The idea that you shouldn't end sentences with a preposition has always been an utter hoax. Regardless of one's esteem for any book or person who taught it to you, it's utterly baseless."

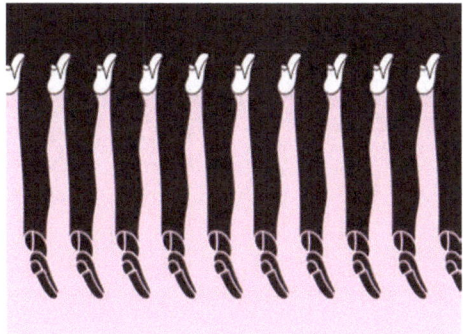

Figure 1.2: Ambiguity (Vecteezy Library)

Here, the ambiguity lies in the interpretation of 'values', which could refer to the input or output of the function. The clarification specifies that it's the input values that are being considered.

———

Place the point at the circle that is inside the square.

Place the point at the center of the circle, which is inside the square.

It's unclear whether the point is to be placed inside the circle or if 'inside the square' modifies where the circle should be.

———

The relationship between the variables changes with the parameter.

As the parameter increases, the direct relationship between the two variables becomes inversely proportional.

The statement is vague about how the relationship changes. The clarification specifies the nature of the change.

Little words II.
Pronouns

The most commonly used pronouns include personal pronouns (he, she, it, they), possessive pronouns (his, her, its, their), and demonstrative pronouns (this, these, that, those).

The primary challenge with pronouns is ambiguity. A pronoun must clearly refer to the specific noun it replaces.

Ensure that the noun referenced by each pronoun is clear.

Figure 2.1: I and you (Vecteezy Library)

It's often helpful to repeat the noun if there's any chance of ambiguity.

Avoid overuse

While pronouns can enhance readability by eliminating repetition, overusing them, especially in complex explanations or proofs, can make your writing difficult to

Supplementary Information The online version contains supplementary material available at https://doi.org/10.1007/978-3-031-65161-8_2.

follow.

Examples:

In geometry, consider a triangle with equal sides. The angles opposite them are also equal.

In geometry, consider a triangle with equal sides AB and AC. The angles opposite these sides, $\angle B$ and $\angle C$, are also equal.

In number theory, consider a sequence of integers where each number is the sum of the two preceding numbers. It begins with 0 and 1. As it progresses, each term is the sum of the previous two terms. For example, the third term is the sum of the first two terms, and the fourth term is the sum of the second and third terms. It continues indefinitely.

In number theory, consider a sequence of integers where each number is the sum of the two preceding numbers. This sequence begins with 0 and 1. As the sequence progresses, each term is the sum of the previous two terms. For example, the third term is the sum of the first two terms, and the fourth term is the sum of the second and third terms. This pattern continues indefinitely.

Use of 'it' and 'this'

These pronouns can be particularly vague. For example, 'This shows' or 'It follows that' should clearly connect to the specific argument being discussed.

After we defined the function, it was clear.

What does 'it' refer to? The clarity of the function, the definition process, or something else?

After we defined the function, its properties became clear.

Now, 'its' clearly refers to 'the function'.

Engaging but formal

We can see that the solution satisfies all the given conditions.

I and we

The phrase 'the royal we' refers to the use of a plural pronoun (such as 'we' or 'us' by a single person, to refer to themselves. Mathematicians use 'we' to suggest that the author and the reader are together exploring concepts, problems, and proofs. See also Section 2 in Halmos's Appendix C.

However, to use 'we' as a modest 'I' is misplaced.

We conjecture that …

I conjecture that …

Tilting at windmills?

Figure 2.2: Don Quijote (Vecteezy Library)

Other classes of pronouns

There are many other classes of pronouns. I list a few.

Personal pronouns
>   Subjective: I, you, he, she, it, we, they;
>
>   Objective: me, you, him, her, it, us, them.

Possessive pronouns
>   Adjective form: my, your, his, her, its, our, their;
>
>   Noun form: mine, yours, his, hers, ours, theirs.

Reflexive pronouns
>   Myself, yourself, himself, herself, itself, ourselves, yourselves, themselves.

Indefinite pronouns
>   Anyone, anything, everybody, somebody, each, few, many, nobody, some, etc.

Relative pronouns
>   Who, whom, whose, which, that.[1]

Demonstrative pronouns
>   This, that, these, those.

Interrogative pronouns
>   Who, whom, whose, which, what.

---

[1] See Section 4.4.

Reciprocal pronouns
    Each other, one another.

Intensive pronouns   (similar to reflexive, emphasizing the subject)

    Myself, yourself, himself, herself, itself, ourselves, yourselves, themselves.

    This list includes the most common English pronouns, but there are more, especially within the indefinite category: 'someone', 'everything', 'anybody', etc.

TOPIC

3

Little words III.
Conjunctions

Figure 3.1: The conjunction 'or' (Vecteezy Library)

Conjunctions are words that join together other words, phrases, clauses, or sentences, such as 'as', 'and', 'but', 'or', 'yet', 'although'.

Supplementary Information The online version contains supplementary material available at https://doi.org/10.1007/978-3-031-65161-8_3.

G. Grätzer, *The Little Book of Math into English*, https://doi.org/10.1007/978-3-031-65161-8_3

11

You're given a set of instructions to select numbers from a list. The instructions are:

Choose numbers that are greater than 10 *and divisible by* 3.

The conjunction 'and' requires that both conditions be met simultaneously. The next instruction:

Choose numbers that are even and odd from the list.

This is incorrect because a number cannot be both even and odd at the same time. The conjunction here connects two mutually exclusive conditions.

Identify numbers that are multiples of 4 *and are less than* 20.

This statement uses a conjunction appropriately, combining two conditions that can simultaneously apply to several numbers.

Little words IV.
so, such, that,...

Figure 4.1: so that, such that (Vecteezy Library)

Supplementary Information The online version contains supplementary material available at https://doi.org/10.1007/978-3-031-65161-8_4.

4.1   So, so that, such that

So is commonly used as a conjunction or adverb to indicate consequence or result.

> *The proof was validated by multiple tests, and* so *the theorem is accepted as true.*

Do not overuse it!

So that is a conjunction that expresses purpose or intention. It might be used to clarify the purpose behind steps in a proof.

> *We introduce this constraint* so that *the system remains stable under all conditions.*

Such that is a phrase that introduces a condition or requirement.

> *Let S be the set of all integers x* such that *$x^2 < 16$.*

In this example, 'such that' introduces the condition $x^2 < 16$, which must be satisfied for $x \in S$.

4.2   This, these, that, those, some

this and these

> Ambiguous:

> *After proving the theorem,* this *becomes evident.*

> Clarified:

> *After we proved the thorem, the correlation between the two variables becomes evident.*

> Vague:

> *These are significant in solving the problem.*

> Specific:

> *These methods are significant in solving the optimization problem.*

that and those

> Vague:

> *That is a well-known result.*

> Clear:

> *That theorem is a well-known result in number theory.*

> General:

> *Those were first introduced in the 20th century.*

> Detailed:

> *Those techniques were first introduced in the field of computational math in the 20th century.*

THIS

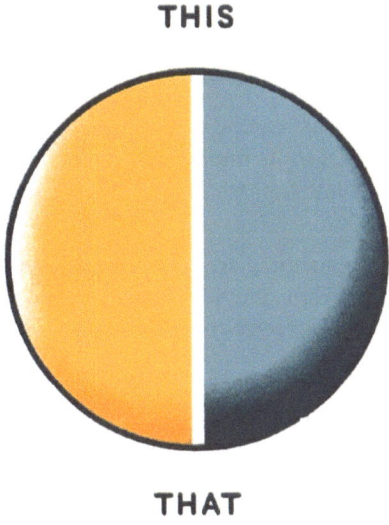

THAT

Figure 4.2: This and that (Vecteezy Library)

some

Vague:

Some *solutions to the equation are real numbers.*

Precise:

Some *solutions to the equation are real, namely, those that are positive.*

General:

Some *of the most fascinating discoveries in math have been accidental.*

Specific:

Some *of the most fascinating discoveries in mathematics, such as the discovery of fractals, have been accidental.*

4.3   *This or that*

The clarity of 'this' versus 'that' isn't always obvious. The reason lies in the context and the author's relationship to the subject matter.

Here's a simplified guide.

Proximity in space or time 'This' is used to refer to something that is physically or temporally (with regard to time) near. For example, in "This book on my desk is my favorite."

'This' points to a book that is close to the author.

'That' refers to something that is further away in space or time. For example, "That mountain in the distance looks daunting." where 'that' is used to talk about a mountain far from the speaker.

Specificity and emphasis 'This' can be used to highlight a specific item or detail, often drawing attention to its importance or relevance at the moment. For instance, "This moment is all we have," emphasizes the importance of the current moment.

'That' can reference something previously mentioned or known, but it is not the immediate focus. For example, "Remember that time we went hiking? That was fun', where 'that' refers back to a specific time mentioned.

Abstract ideas When referring to concepts, ideas, or experiences, both 'this' and 'that' can be used to convey proximity or distance in terms of emotional or psychological connection, not just physical or temporal. "This idea seems promising" might suggest a new or currently discussed concept, whereas "That old belief of ours turned out to be false" points to a more distant or past concept.

More examples

To solve this equation, first distribute the values. This step is crucial for simplifying the equation. 😊

———

After distributing the values as shown in that step, move on to combining like terms. 😊

———

In this context, 'this' refers to a step or part of the process currently being explained or focused on, making it feel more immediate. 'That' refers to a step that has been previously mentioned or is considered already understood or completed, placing it at a conceptual distance.

To prove this theorem, we first assume that this is true... 😊

That assumption we made earlier? It allows us to conclude... 😊

'This' is used to focus on the current step or element of the proof being developed or assumed. 'That' references an earlier step or assumption, suggesting its importance in reaching the conclusion.

———

Let's consider a function $f(x)$ defined on the interval $[0, 1]$. If $f(x)$ has a maximum at $x = 0.5$, then this maximum value is significant because of that property. 😞

It's unclear what "this maximum value" refers to specifically (is it the value of $f(0.5)$ or the fact that the maximum occurs at 0.5?), and "that property" is

Figure 4.3: The function $f(x)$ with a maximum (Vecteezy Library)

vague (does it refer to the function being defined on $[0, 1]$, or to the function having a maximum?).

Let's consider a function $f(x)$ defined on the interval $[0, 1]$. If $f(x)$ has a maximum at $x = 0.5$, then the maximum value $f(0.5)$ is significant because of its defining property as the highest point on the interval. 😊

Given two parallel lines cut by a transversal, this creates corresponding angles that are equal. This means that the lines will never intersect. 🙁

The first 'this' vaguely refers to the scenario of two parallel lines being cut by a transversal, but it's not clearly connected to the angles. The second 'this' refers to the earlier angle equality but it's not explicitly tied to the logic leading to the conclusion about lines not intersecting.

The intersection of two parallel lines with a transversal creates corresponding angles that are equal. And the equality of these two angles ensures that the two lines remain parallel and will never intersect. 😊

4.4   Which and that

Restrictive vs. non-restrictive clauses

A 'restrictive clause' provides information that is crucial to the meaning of the sentence. If a restrictive clause is removed, the meaning of the sentence changes significantly. A restrictive clause is not set off by commas and often uses 'that'.

A 'non-restrictive clause' adds extra information to the sentence that is not critical to understanding it. Removing a non-restrictive clause does not change the meaning of the sentence. These clauses are usually set off by commas and often use 'which'. (Halmos agrees, see Section C.12.)

Using 'that'

Restrictive clause:

The theorem that *revolutionized quantum mechanics has been widely debated.*
😊

Here, "that revolutionized quantum mechanics" is a restrictive clause because it specifies the particular theorem being discussed.

Using 'which'

Non-restrictive clause:

Schrödinger's Cat, which *is a thought experiment, illustrates quantum superposition.* 😊

"which is a thought experiment" is a non-restrictive clause. It adds information about Schrödinger's Cat but is not necessary to understand the main point.

Comma as a cue

If you're adding information that doesn't change the meaning of a sentence, use 'which' and set off the clause with commas. If the information is essential and removing it would alter the meaning of the sentence, use 'that' without commas.

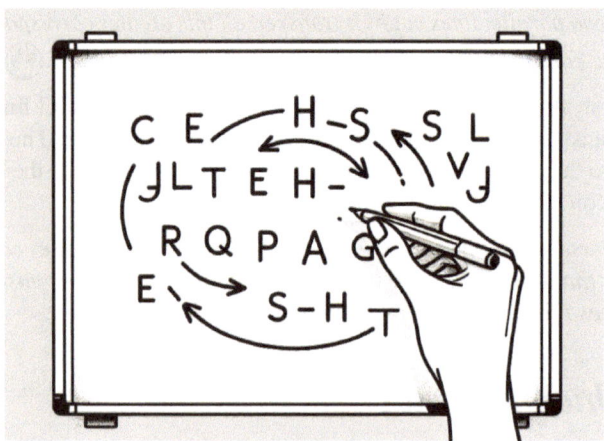

Figure 4.4: Rearrange the sentence (Vecteezy Library)

Rearrange the sentence

If the sentence becomes difficult to read with the 'that' clause (restrictive clause), try to restructure it.

Before rearranging it:

The theorem shows that, for every epsilon greater than zero, there exists a delta. 😣

After rearranging it:

For every epsilon greater than zero, the theorem shows that there exists a delta.

According to the theorem, for every epsilon greater than zero, there exists a delta [such that what]

If the inserted remark is an aside or provides clarification, consider using parentheses.

The proof demonstrates that (given the initial assumptions) the variable cannot exceed the boundary.

When a sentence becomes too complex, consider breaking the sentence into two or more sentences.

Before simplification:

The study confirms that, despite the initial hypothesis suggesting otherwise, the data are consistent with the theoretical model.

After simplification:

The study confirms that the data are consistent with the theoretical model, despite the initial hypothesis suggesting otherwise.

Little words V.
Like, such as

The distinction between 'like' and 'such as' is subtle yet significant. 'Like' draws comparisons or indicates similarity.

Functions like the one we have discussed can be highly complex. 😊

This indicates similarity, not specific examples.

Equations like this often require innovative solutions. 😊

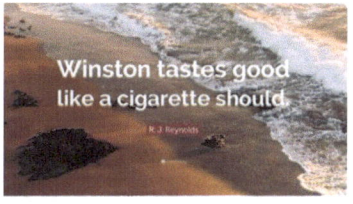

Figure 5.1: Winston tastes good like a cigarette should

Suggests a general category of equations.

'Such as' introduces specific examples.

Functions such as polynomials and exponentials have distinct characteristics. 😊

This lists specific examples of functions.

Equations such as Euler's identity embody the beauty of mathematics. 😊

Native English speakers are at a disadvantage. They are inundated by the echoes of the Winston cigarette advertisement, and 'like' sounds to them like 'as'. Wiki has a long discussion: "Grammar controversy".

During the campaign's long run in the media, many criticized the slogan as

Supplementary Information The online version contains supplementary material available at https://doi.org/10.1007/978-3-031-65161-8_5.

G. Grätzer, *The Little Book of Math into English*, https://doi.org/10.1007/978-3-031-65161-8_5

grammatically incorrect, asserting that it should say, "Winston tastes good as a cigarette should." Ogden Nash, in The New Yorker, published a poem that ran "Like goes Madison Avenue, like so goes the nation." Walter Cronkite, then hosting The Morning Show, refused to say the line as written, and an announcer was used instead.

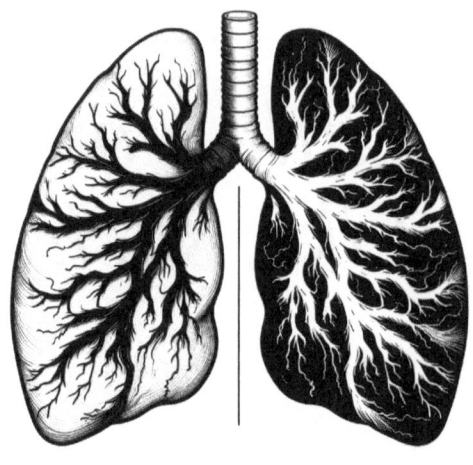

Figure 5.2: Not so good (Vecteezy Library)

Little words VI. Either, or, both

Either

'Either' signifies one out of two mutually exclusive options.

Examples:

A proof is either deductive or inductive, depending on its logical structure.

_____

In computational complexity, a problem is classified as either NP-hard or not.

Here, 'either' introduces scenarios where there are two distinct, mutually exclusive categories.

Or

'Or' links alternatives and does not signify an exclusive relationship.

Exclusive use (implied by context):

The shape of a fundamental neighborhood is either circular or square.

Inclusive use:

Your study may focus on theory or application, but not both.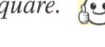

Supplementary Information The online version contains supplementary material available at https://doi.org/10.1007/978-3-031-65161-8_6.

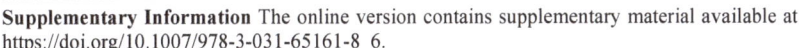

If 'or' is used, consider adding a brief clarification if there's any risk of ambiguity.

Both

'Both' is used to indicate that two conditions or statements are applicable simultaneously.

Examples:

A well-written math paper is both clear and concise.

―――――

An algorithm must be efficient in terms of both time and space to be considered optimal.

Either... or

An 'either... or' statement (or 'either... or else' statement) presents two mutually exclusive options, implying that if one is true, the other must be false.

Either the sum of 5 and 7 is equal to 12, or the product of 2 and 4 is equal to 9.

In this sentence, you are presented with two statements; the first statement is true.

Little marks I. Punctuation

7.1 Background

The comma's origins can be traced back to ancient Greece, where Aristophanes of Byzantium, a scholar and librarian in Alexandria, introduced a system of dots (distinctions) around the 3rd century BCE to aid readers in understanding the rhythm and pauses in sentences. These dots were placed at different text heights to represent short, medium, and long pauses; see J. E. Sandy [18].

Figure 7.1: Aristophanes (Vecteezy Library)

Supplementary Information The online version contains supplementary material available at https://doi.org/10.1007/978-3-031-65161-8_7.

7.2   The two approaches

Elocution method

Aristophanes' method evolved into the 'elocution method' of placing commas, emphasizing the spoken aspect of language, where commas represent pauses, intonation changes, or breaths taken by the speaker. The elocution method is flexible, relying on the speaker's judgment.

The grammar approach

In this approach, the use of commas is rule-based, focusing on the written language. It follows specific guidelines for comma placement (separating items in a list, for instance). This approach prioritizes clarity.

Conflict

Conflicts arise when the strict application of grammatical rules does not align with natural speech patterns.

Grammatical approach:

In the morning, we will analyze the data, and, after lunch, we will discuss the results. 😊

Or simply,

In the morning we will analyze the data, and after lunch we will discuss the result. 😊

Elocution method:

In the morning, we will analyze the data and, after lunch we will discuss the results. 😊

A speaker might naturally pause after 'analyze the data' without the need for a comma before 'and' or might not pause at all after 'lunch'.

The real confusion

Native English speakers start learning punctuation by reading as children. Non-native English speakers learn punctuation by reading graduate books and research articles. Neither group analyzes the examples remembered, whether they follow the grammatical method or the elocution method. The result: a mixed approach.

Confusion begets confusion.

Figure 7.2: Confusion begets confusion (Vecteezy Library)

7.3   Commas between multiple adjectives

When you have two or more adjectives that modify the same noun, the commas (or lack thereof) can subtly change the meaning of the sentence. The general rule is that if the adjectives are 'coordinate' (meaning they independently modify the noun), they should be separated by a comma. If they are 'cumulative' (meaning they build on one another to modify the noun), no comma is needed.

Coordinate Adjectives

'Coordinate adjectives' are adjectives that can be rearranged without changing the meaning of the sentence, and you can place 'and' between them without altering the sentence's sense. In these cases, use a comma.

Examples:

The loud obnoxious laugh echoed through the room.

The loud, obnoxious laugh echoed through the room.

———

The quadratic flat equations puzzled the students.

The quadratic, flat equations puzzled the students.

In this example,"quadratic" and "exponential" are coordinate adjectives describing "equations". You can test this by adding "an" between them or changing their order without losing the meaning:

The exponential and quadratic equations puzzled the students.

Cumulative Adjectives

Cumulative adjectives do not independently modify the noun. Instead, each adjective modifies the combination of the noun and the following adjective(s), creating a specific meaning. Therefore, you don't use a comma.

Examples:

The dark chocolate cake melted in her mouth.

———

The complex polynomial function graph twisted across the screen.

In this example, "complex" modifies not just "function" but "polynomial function", indicating a specific type of function. You can't rearrange these adjectives or insert "and" without altering the meaning.

7.4   The role of the punctuation marks

Commas

Use to separate items in a list, after introductory phrases, between adjectives, before conjunctions in compound sentences, and to set off non-essential information.

Figure 7.3: Comma-itis—treat it with SA14-14-2 (Vecteezy Library)

Semicolons

Use to connect closely related independent clauses or to separate items in a list where the items themselves contain commas.

Colons

Use to introduce a list, quote, explanation, or elaboration that follows a complete sentence.

Parentheses

Use sparingly to include non-essential but informative details or side notes.

7.5   Using commas

Commas for listing Commas are used to separate items in a list.

For example:

In our research, we explored topology algebraic structures and computational complexity. 😩

In our research, we explored topology, algebraic structures, and computational complexity. 😊

This is an example of an 'Oxford comma', discussed in the next section.

Commas and clauses Commas are used to separate independent clauses linked by conjunctions.

The professor taught the concept in class however the students found it confusing. 😩

The professor taught the concept in class, however, the students found it confusing.

Commas for precision Commas can clarify meaning and prevent misinterpretation:

The professor noted the solution on the board was incorrect.

The professor noted, the solution on the board was incorrect.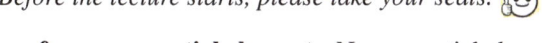

Commas with introductory elements Use commas after introductory words, phrases, or clauses that precede the main clause.

Before the lecture starts please take your seats.

Before the lecture starts, please take your seats.

Commas for non-essential elements Non-essential elements add information to a sentence but do not change the overall meaning.

The theorem that was proved last year has already been applied in various fields.

The theorem, which was proved last year, has already been applied in various fields.

Do not overuse Unnecessary commas can interrupt the flow of thought.

The solution, was elegant, and simple.

The solution was elegant and simple.

Misuse of semicolons Do not overuse semicolons where commas would suffice.

My favorite algebraic systems are lattices; groups; rings; and semigroups.

My favorite algebraic systems are lattices, groups, rings, and semigroups.

———

She opened the article; read the first page; and then closed the volume.

She opened the article, read the first page, and then closed the volume.

———

He was late for the meeting; because his car broke down.

He was late for the meeting because his car broke down.

7.6 Oxford comma

The Oxford comma is the final comma in a list of items that comes before the conjunction (usually 'and' or 'or'). For example, in the list "rings, groups, and semigroups," the Oxford comma is the one after "groups".

The use of the Oxford comma is a stylistic choice. Mathematicians should rejoice. Finally, a rule of grammar that is easy to use, with no ambiguity!

Examples:

I dedicate this book to my parents,
J. Frase, and K. Frase. 😊

Without the Oxford comma, this could be misread: the author's parents are J. Frase and K. Frase.

———

Without Oxford comma:

In today's math class, we'll cover addition, subtraction and multiplication and division. 😩

With Oxford comma:

In today's math class, we'll cover addition and subtraction, multiplication and division. 😊

The Oxford comma makes it clear that multiplication and division are separate topics, not a combined one.

Without Oxford comma:

The competition includes algebra, number theory, geometry and trigonometry challenges. 😩

With Oxford comma:

The competition includes algebra, number theory, geometry, and trigonometry challenges. 😊

Here, the Oxford comma shows that geometry and trigonometry are distinct challenges.

Figure 7.4:   Oxford comma (Vecteezy Library)

———

Without Oxford comma:

For our study session, please bring pencils, erasers, notebooks and calculators. 😩

With Oxford comma:

For our study session, please bring pencils, erasers, notebooks, and calculators. 😊

This clarifies that notebooks and calculators are separate items to bring to the study session.

———

Without Oxford comma:

Please complete the exercises on pages 12, 15, 17 and 19. 😩

With Oxford comma:

Please complete the exercises on pages 12, 15, 17, and 19.

The Oxford comma helps differentiate between the page numbers of the exercises assigned as homework.

Style guides vary in their recommendations. The Chicago Manual of Style [5] and the Oxford University Press Style Guide [8] recommend using the Oxford comma.

An entertaining story   A famous legal case[1] involved the lack of an Oxford comma in a law regarding overtime pay, leading to a dispute over whether certain activities were exempt from overtime.

The sentence in question listed overtime exemptions for "packing for shipment or distribution" without a comma before "or distribution", leading to a lawsuit over whether the law meant "packing for shipment" and "distribution" as separate exemptions or "packing for shipment or distribution" as a single activity.

The court ruled in favor of the workers, emphasizing the importance of clarity that the Oxford comma could have provided.

---

[1] United States Court of Appeals For the First Circuit, No. 16-1901, March 13, 2017.

Little marks II. Hyphens

8.1 When to use hyphens

Hyphens play many roles.

Compound modifiers

Figure 8.1: Hyphen (Vecteezy Library)

When two or more words are used together as a single adjective before a noun, they should be hyphenated.

Examples:

Without hyphen:

A well known problem in topology is... 😫

This could be misread, the problem is well and known in topology.

With hyphen, the correct way:

Supplementary Information The online version contains supplementary material available at https://doi.org/10.1007/978-3-031-65161-8_8.

A well-known problem in topology is... 😊

But contrast this with:

This topological problem is well known. 😊

There no hyphen here, since "well" modifies "is known".

Avoiding ambiguity

Examples of ambiguity:

A 'small animal habitat' could mean a habitat that is small for animals. 😖

Clear:

A 'small-animal habitat' clarifies that the habitat is for small animals. 😊

Prefixes and suffixes

Hyphens are used in math to improve clarity, especially when dealing with prefixes and suffixes.

Re-sign vs. Resign In math, you might need to 're-sign' a number, meaning to assign it a new sign. Without the hyphen, 'resign' means quitting.

After reviewing the calculations, the mathematician decided to re-sign the number, changing -8 to +8 to correct the error. 😊

Frustrated with the ongoing challenges, the lead researcher decided to resign from the project. 😊

Re-cover vs. Recover 'Re-cover' means to cover something again. 'Recover' means to get back to a normal state, such as a function recovering its original value.

The team decided to re-cover the telescope with its protective case after observing the solar eclipse. 😊

The function quickly began to recover its original form once the perturbing forces were removed. 😊

Like For instance, 'shell-like' clarifies that something is similar to a shell.

The newly discovered exoplanet has a shell-like structure, indicating layered atmospheric composition. 😊

In chemistry class, we learned that electrons move in defined paths known as shells around the nucleus. 😊

Many prefixes (e.g., 'non', 'multi', 'micro', 'semi') do not typically require a hyphen.

Examples:

'Nonlinear'.

No hyphen is necessary.

"Semisimple Lie algebra"

No hyphen is necessary.

'Semi-independent'.

Hyphen used for clarity, 'semiindependent' is difficult to read.

Figure 8.2: 20-year-old blond (Vecteezy Library)

Numbers and fractions

In math writing, hyphens are often used in compound numbers and fractions when they are written out in words.

For example, 'twenty-three' is used in compound numbers, and 'two-thirds' is used in fractions.

Avoiding hyphens with adverbs ending in 'ly'

Adverbs that end in 'ly' and modify an adjective do not need a hyphen when used together before a noun, because the 'ly' ending makes the relationship clear.

Example:

A rapidly converging series.

A rapidly-converging series.

8.2 Examples of hyphenated words

I now present some examples of commonly used hyphenated terms in math.

Compound modifiers

Well-known theorem

Long-term project

Two-dimensional space

Finite-dimensional vector
space

Non-linear equation

Semi-simple Lie algebra

Well-defined function

Fixed-point theorem

Single-variable calculus

Also acceptable: nonlinear, semisimple.

Prefixes and special terms

Re-entry	Half-life
Co-author	Quarter-turn
Pre-calculus	Step-function
Multi-variable	Third-degree polynomial
Cross-section	

Also acceptable: precalculus and multivariable.

Numbers and fractions

Twenty-first century	One-half
Two-thirds majority	Three-quarters

Compound adjectives

Hyphens are often used in compound adjectives before a noun.

When using prefixes like 'non-', multi-', 'semi-', and 'pre-', hyphens can clarify meaning, especially when the prefix precedes a capital letter or a numeral.

Non-: Used to indicate negation or absence

Non-Euclidean Geometry	Multi-digit Numbers
Non-zero	
Multi-variable Calculus	Semi-simple Lie Algebra

Also acceptable: noneuclidean, nonzero, multivariable, semisimple (standard).

Pre-: Indicates before
Pre-calculus   Pre-image
Also acceptable: preimage (standard), precalculus.

8.3   A list of hyphenated words

Barbara Beeton worked for decades for the AMS and coedited the TUGboat, The Communications of the TeX Users Group.

In these capacities, she compiled lists of words that LaTeX does not hyphenate properly. The built-in hyphenation patterns of LaTeX usually do their jobs properly but may stumble on a simple word like 'database'.

This hyphenation problem is quite different from the hyphenations we discussed before on the topic. There, we discussed hyphenations that should always be present:, such as "a well-known problem". The hyphenations we now discuss are intended to show where to break the words when they must be broken between the end of one line and the beginning of the next. (My proud contribution to her list: Chicago.)

Barbara Beeton [1] compiled a long list of these exceptions. LaTeX will correctly hyphenate these with the \hyphenation command in the preamble of the article. So if you want 'database' properly hyphenated, add the command

\hyphenation{data-base}.

Figure 8.3: Barbara Beeton with her favorite hat

Here is a small sample from Beeton's list. The first column is "as is", the way LaTeX understands the hyphenation pattern. The second column is "as it should be", the way LaTeX should understand the hyphenation pattern.

`academy(ies)`	acad-e-my(ies)
`ac-cusative`	ac-cu-sa-tive
`acronym`	acro-nym
`acupunc-ture(ist)`	acu-punc-ture(-ist)
`adamant`	ad-a-mant
`addable`	add-a-ble
`ad-di-ble`	add-i-ble
`adrenaline`	adren-a-line
`aerospace`	aero-space
`af-terthought`	af-ter-thought
`agronomist`	agron-o-mist
`al-ge-braically`	al-ge-bra-i-cal-ly
`al-manac`	al-ma-nac
`al-manack`	al-ma-nack
`anachro-nism(tic)`	anach-ro-nism(-tic)
`analect`	an-a-lect
`anal-yse(d)`	an-a-lyse(d)
`anal-y-ses`	analy-ses
`anal-y-sis`	analy-sis
`aneurysm`	an-eu-rysm
`aneurys-mal`	an-eu-rys-mal
`anisotropic`	an-iso-trop-ic
`anisotrop-i-cally`	an-iso-trop-i-cal-ly
`anisotropism`	an-isot-ro-pism
`anisotropy`	an-isot-ropy
`an-niver-sary(ies)`	an-ni-ver-sary(ies)
`anomaly(ies)`	anom-a-ly(ies)
`anonymity`	an-o-nym-i-ty
`anony-mous(ly)`	anon-y-mous(ly)
`an-tibi-otic`	anti-bi-ot-ic
`an-tideriva-tive`	anti-deriv-a-tive
`an-ti-holo-mor-phic`	anti-holo-mor-phic
`anti-nomy(ies)`	an-tin-o-my(ies)
`antin-u-clear`	anti-nu-clear
`antin-u-cleon`	anti-nu-cle-on
`an-tipodes`	an-tip-o-des
`an-tirev-o-lu-tion-ary`	anti-rev-o-lu-tion-ary
`ape-ri-odic`	a-peri-odic
`apotheoses`	apoth-e-o-ses
`apotheo-sis`	apoth-e-o-sis
`ap-pen-dices`	ap-pen-di-ces
`ap-pendix`	ap-pen-dix
`archimedean`	ar-chi-me-dean
`archipelago`	ar-chi-pel-ago
`archipela-gos`	ar-chi-pel-a-gos
`archive`	ar-chive
`archiv-ing`	ar-chiv-ing

Little marks III. Dashes

The two dashes, the en-dash and the em-dash, are older siblings of the hyphen. In LaTeX, type

```
--
```

for the en-dash and type

```
---
```

for the em-dash.

The en-dash (–)

The en-dash is shorter than the em-dash but longer than the hyphen. It is most commonly used to indicate ranges or connections.

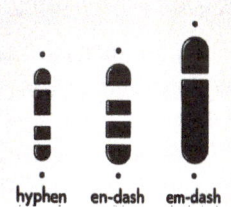

Figure 9.1: Dashes
(Vecteezy Library)

1. To indicate ranges   of numbers, dates, or times, suggesting continuity between the values.

> *The conference runs June 1–5.* 😊
> Not June 1-5. Even better:
> *The conference runs from June 1 to 5.* 😊

2. To connect numbers, words, or names   The en-dash can also connect items that are related or in contrast to each other, including scores or connections between

---

Supplementary Information The online version contains supplementary material available at https://doi.org/10.1007/978-3-031-65161-8_9.

places or names.

The London–Paris train.

———

The Euler–Lagrange equation.

———

The method was effective in cases of high variance–low bias, but not the opposite.

———

Our research contrasts the traditional–modern approaches to calculus.

———

The report details the gap in income distribution: top 1%–bottom 99%.

The em-dash (—)

The em-dash is longer than the en-dash and is used in several ways:

1. To add emphasis or an interruption   Em-dashes can replace commas, parentheses, or colons to add emphasis or introduce an interruption, which is particularly
useful in math writing to highlight important points or definitions.

The function $f(x)$—defined on the interval $[0, 1]$—is continuous.

2. To indicate parenthetical statements   Similar to
parentheses, em-dashes can enclose additional information.

*The theorem—one of the most pivotal in number
theory—was proved in 1994.*

3. To set off lists   An em-dash can introduce a list
or a restatement, offering a stylistic alternative to the
colon.

*Three primary colors—red, blue, and yellow—
form the basis of the color wheel.*

No spaces before or after an em-dash or an en-
dash.

Figure 9.2: Color wheel

Em-dashes, —, and parentheses, (), serve different roles in punctuation, especially when it comes to emphasizing or de-emphasizing parts of a sentence.

When to use em-dashes

Em-dashes are used to create a strong break in the structure of a sentence and are
excellent for adding emphasis to the enclosed text.

Example:

*The formula to find the area of a circle is $A = \pi r^2$, where r is the radius of the
circle. It is utilized...*

More emphatic:

The formula to find the area of a circle—$A = \pi r^2$—is utilized...

In this example, the em-dashes draw attention to the formula itself, making it the focal point of the sentence.

Easier to read:

The formula to find the area of a circle ($A = \pi r^2$) is utilized...

Parentheses for de-emphasis

Parentheses are used to include additional information that is considered less important, providing extra details without detracting from the main point of the sentence. The information inside the parentheses is often seen as an aside or extra detail that is not essential to the main message.

Example:

The solution to the quadratic equation is given by the formula

$$x = \frac{-b \pm \sqrt{b^2 - 4ac}}{2a},$$

where a, b, and c are coefficients of the equation.

With de-emphasis using parentheses:

The solution to the quadratic equation is given by the formula

$$x = \frac{-b \pm \sqrt{b^2 - 4ac}}{2a}$$

(where a, b, and c are coefficients of the equation).

Even better:

The solution to the quadratic equation is given by the formula

$$x = \frac{-b \pm \sqrt{b^2 - 4ac}}{2a}$$

(a, b, and c are coefficients of the equation).

In this example, the parentheses are used to de-emphasize the explanation about *a*, *b*, and *c*, suggesting that this detail is supplementary rather than central to the understanding of the quadratic formula.

TOPIC

10

Little transitions. Short version

Transition words connect two sentences to add structure. Here is an example using the most popular transition word: 'so'.

> *The sequence is bounded.*
> **So** *it is monotonic and it must*
> *converge.*

Transition words come in six categories.

Cause and effect

Show the relationship between an action or event and its outcome.

Figure 10.1: Transitions
(Vecteezy Library)

> Consequently, *increasing the dimensions of a rectangle will increase its area.*

Comparison

Highlight similarities between two or more ideas or objects.

> **Similarly**, *the process of solving linear equations is akin to finding the intersec-*

Supplementary Information The online version contains supplementary material available at https://doi.org/10.1007/978-3-031-65161-8_10.

tion point of lines on a graph. 😊

Contrast

These words point out differences between two or more ideas or objects.

On the other hand, *while addition involves combining quantities, subtraction involves finding the difference between quantities.* 😊

On the one hand, *Ellie finds the elegance of prime numbers captivating;* **on the other hand,** *she struggles to see their application in her everyday life.* 😊

Sequence/order

Indicate a chronological or logical sequence of events or ideas.

First, *we identify the known variables.* **Next,** *we establish a relationship between them.* **Finally,** *we solve for the unknown.* 😊

Example/emphasis

Introduce examples or highlight important points or ideas.

For instance, *in the context of probability, the outcome of rolling a six-sided die is a prime example of a random event.* 😊

Figure 10.2: Dice (Vecteezy Library)

Conclusion/summary

Signal the end of a discussion and often introduce a summary or conclusion.

In summary, *the fundamental theorem of algebra guarantees that every non-constant polynomial equation has at least one complex root.* 😊

We'll discuss these in Topic 11.

Choose the transitions carefully so that they accurately reflect the relationship between the ideas you're connecting.

Vary your transition words and phrases. A paragraph with many 'so's is soo sloppy.

Do not put commas after the short ones. 'So,' looks pretentious. Follow the rule: no commas if the transition word is fewer than four characters.

Little transitions. Longer version

There are six types of transition words (and phrases).

1. Addition

Adds information or ideas.

Figure 11.1: Transition (Vecteezy Library)

Supplementary Information The online version contains supplementary material available at
https://doi.org/10.1007/978-3-031-65161-8_11.

G. Grätzer, *The Little Book of Math into English*, https://doi.org/10.1007/978-3-031-65161-8_11

- furthermore (emphasizes the strength of the argument)
    Data sets often contain outliers. **Furthermore,** *these can significantly affect the mean.*

- moreover (emphasizes the significance of the argument)
    Primes are infinite in number. **Moreover,** *they are the building blocks of all natural numbers.*

- in addition (more neutral term)
    Graph theory is used in computer science. **In addition,** *it's crucial for network analysis.*

- also
    Triangles have three sides. **Also,** *their interior angles add up to* 180 *degrees.*

2. Comparison and contrast

Signals differences—or similarities—between ideas.

- however (introduces a statement that contrasts with an earlier statement)
    Squares have equal sides. **However,** *rectangles do not.*

- on the other hand and on the other hand (considering an alternative perspective)
    Circles have no corners. **On the other hand,** *polygons do.*

- by contrast and in contrast (directly compare two or more items; stronger than 'however' in emphasizing difference)—also by contrast
    Multiplication speeds up addition. **By contrast,** *division breaks down multiplication.*

- similarly (indicates likeness)
    Even numbers are divisible by 2. **Similarly,** *multiples of* 5 *end in* 0 *or* 5.

Figure 11.2: Comparison and contrast (Vecteezy Library)

3. Cause and effect

Shows the relationship: condition and outcome.

- therefore (direct conclusion) *All angles in a triangle add up to 180 degrees. **Therefore,** the third angle must be 60 degrees if two angles are 60 degrees each.* 😊

- thus (similar to 'therefore', but implies a softer transition)
    *A square has all sides equal. **Thus,** if one side is 4 units, the perimeter is 16 units.* 😊

- as a result (explicitly attributes a consequence of the preceding facts)
    *A line segment bisected at its midpoint divides it into two equal parts. **As a result,** each segment is half the length of the original segment.* 😊

- consequently (a cause-and-effect relationship)
    *Two lines parallel to a third line are parallel to each other. **Consequently,** they never intersect.* 😊

4. Sequence and order

Organizes information chronologically or logically.

- the words 'first', 'next', 'then', and 'finally' are used to organize steps in an argument.
    ***First,** isolate the variable on one side of the equation. **Next,** add or subtract terms on both sides to balance the equation. **Then,** multiply or divide to solve for the variable. **Finally,** check your solution by substituting the variable back into the original equation.* 😊

- The sequence of words: 'initially', 'subsequently', ' meanwhile', 'finally' indicate the steps in a process.
    ***Initially,** we assume that all sides of a given triangle are equal in length. **Subsequently,** we apply the Pythagorean theorem to prove that the triangle is right-angled. **Meanwhile,** we also demonstrate that the angles opposite the equal sides are equal. **Finally,** by combining these findings, we conclude that the triangle is equilateral and right-angled.* 😊

- 'before', 'after', 'during' signify events relative to each other in time.
    *Check for any simplifications that can be made **before** attempting to solve an algebraic equation.* 😊
    ***After** completing the division of two numbers, round the answer to the nearest tenth if necessary.* 😊

    ———

    ***During** the process of graphing a linear equation, carefully plot each point to ensure the line is straight.* 😊

- 'Next', 'then', 'later' indicate the immediate sequence or logical progression in actions, steps, or events. *To solve a word problem, read through the problem carefully.* **Next***, identify the variables and write down what you are asked to find.* 😊

————

List all the prime numbers you can find below 20. **Then***, use them to find the prime factors of 60.* 😊

————

Figure 11.3: Sequence and order (Vecteezy Library)

Start by drawing the shape on a piece of graph paper. **Later***, you will measure each side to determine the perimeter.* 😊

- 'meanwhile', 'concurrently' describe simultaneous events.

  One student calculates the area of a square; **meanwhile***, another measures its perimeter.* 😊

————

Two teams work on solving different algebraic equations; **concurrently***, they discover shared principles that apply to both.* 😊

5. Emphasis

Words (and the phrase) such as 'indeed', 'in fact', and 'notably' highlight key points.

- indeed (strengthening the statement that precedes it)

- in fact (introduces a piece of evidence or an example that supports the claim)

- most importantly (the most significant point)

  Indeed*, the discovery of calculus revolutionized how we understand change and motion in the universe.* 😊

In fact, *math is the language in which physics describes the universe.*

Most importantly, *critical thinking in math fosters problem-solving skills applicable in everyday life.*

6. Summary and conclusion

- in summary (introduces a summary)

- to sum up (similar to the previous one)

- in conclusion (a strong closing statement or a concise summary of findings)

   In summary, *prime numbers serve as the fundamental building blocks of the integers, a statement that underscores their critical importance in number theory.*

   ———

   To sum up, *Euler's formula connects the worlds of geometry, algebra, and trigonometry elegantly.*

   ———

   To sum up, *Euler's formula elegantly connects the worlds of geometry, algebra, and trigonometry.*

   ———

   In conclusion, *the concept of infinity is not just philosophical but deeply embedded in calculus and analysis.*

The transition words I listed in the categories are examples, not complete lists. To illustrate this, here are the *common* transition words and phrases in the Addition category:

and	additionally
also	again
moreover	too
furthermore	plus
in addition	indeed
besides	in fact
likewise	of course
as well as	then
not only... but also	equally important
equally	furthermore
similarly	more importantly

Little INTROS

An INTRO is a word (or two) INTROducing a formula. Your work is much easier to read if you use plenty of INTROs. Instead of

> *L is distributive.*

use the INTRO 'lattice', so you write

> *The lattice L is distributive.*

This is much better. It gently reminds the reader that L is a lattice. It also follows the old rule: Do not start a sentence with a formula or symbol.

Figure 12.1: Without INTRO, with INTRO (Vecteezy Library)

Supplementary Information The online version contains supplementary material available at https://doi.org/10.1007/978-3-031-65161-8_12.

More examples

What an awful sentence:

Since $Z_{\vec{B}} \cap A_i = \emptyset$, $F_{\vec{B}}^{(r)}$ is a copy of $F^{(r)}$. 🙁

How much easier it is to read:

Since the sets $Z_{\vec{B}}$ and A_i are disjoint, the set $F_{\vec{B}}^{(r)}$ is a copy of the set $F^{(r)}$. 🙂

Since the formula for disjointness is familiar, we can write more briefly:

Since $Z_{\vec{B}} \cap A_i = \emptyset$, the set $F_{\vec{B}}^{(r)}$ is a copy of the set $F^{(r)}$. 🙂

———

We proved that $\gamma(B_k)$ is the smallest $n \in \mathbb{N}_0$ such that $k \leq \binom{n}{\lfloor n/2 \rfloor}$. 🙁

Use some INTROS:

We proved that $\gamma(B_k)$ is the smallest natural number n such that the formula $k \leq \binom{n}{\lfloor n/2 \rfloor}$ holds. 🙂

Display the tall formula:

We proved that $\gamma(B_k)$ is the smallest natural number n such that the following formula holds:

$$k \leq \binom{n}{\lfloor n/2 \rfloor}.$$ 🙂

———

Keeping r fixed, $f_r(n)$ asymptotically equals $\sigma(F^{(r)}, B_n)$; that is, we have that $\lim_{n \to \infty} \sigma(F^{(r)}, B_n)/f_r(n) = 1.$ 🙁

We throw in a few INTROS:

If we keep the natural number r fixed, the function $f_r(n)$ asymptotically equals the function $\sigma(F^{(r)}, B_n)$; that is, the following formula holds:

$$\lim_{n \to \infty} \sigma(F^{(r)}, B_n)/f_r(n) = 1.$$ 🙂

More advanced examples

An INTRO may clarify how a structure is viewed. For instance, the ideal lattice of a lattice can be viewed (among others) as

1. a lattice,

2. an ideal lattice,

3. a closure system.

So

The lattice L is distributive. 🙂

is excellent because 'distributive' is a lattice property.

In the next version, the INTRO and the property do not agree:

The ideal lattice Id L is distributive. 🙁

because *distributive* is not an ideal lattice property.

Accordingly, use the following form:

The lattice Id L is distributive. 🙂

Write
  The closure system Id *L is algebraic.*
because *algebraic* is a property of the ideal lattice as a closure system.

Note

Adding an INTRO is context dependent. Our first example, change
  L is distributive.
to
  The lattice L is distributive.
is not required if *L* has already been identified as a lattice recently.

————

How many INTROs to add? Too many, and the sentence might become too long
and repetitive. Use your judgement.

Little run.
Run-on
sentences

You have just finished the long proof of a theorem and write:

So the lattice L is distributive and the ideal lattice Id *L is finite, concluding the proof.*

I am on a roll... I make this mistake all the time. This is a common example of a 'run-on sentence'.

Run-on sentences occur when two or more 'independent clauses' (clauses that can stand alone as sentences) are connected improperly without appropriate punctuation or conjunctions. Run-on sentences often make the text hard to follow and can confuse the reader.

Figure 13.1: Run-on sentences (Vecteezy Library)

Traditional Run-on Sentence:

I love to write, it is my passion, I spend hours every day at my desk writing stories, none of them are published yet.

It is easy to correct:

I love to write; it is my passion. I spend hours every day at my desk writing

Supplementary Information The online version contains supplementary material available at https://doi.org/10.1007/978-3-031-65161-8_13.

stories, but none of them are published yet.

Improper sequence without separation:

$2 + 2$ *is* 4 4×3 *is* 12 $12/4$ *is* 3

$2 + 2$ *is* 4, 4×3 *is* 12, $12/4$ *is* 3

Equations strung together:

$x = 5$ $y = x + 2 = 7$ $z = y \times 2 = 14$

$x = 5$, $y = x + 2 = 7$, $z = y \times 2 = 14$

Concepts without proper segmentation:

The Pythagorean theorem states a^2 plus b^2 equals c^2 this is used to calculate the hypotenuse of a right triangle the area of a triangle is one-half base times height.

The Pythagorean theorem states a^2 plus b^2 equals c^2. This is used to calculate the hypotenuse of a right triangle. The area of a triangle is one-half base times height.

To avoid run-on sentences in writing, you can:

- Use punctuation such as periods (.) or semicolons (;) to separate independent clauses.

- Use coordinating conjunctions (for, and, nor, but, or, yet, so) with a comma to connect independent clauses.

- Break complex ideas into simpler, more digestible sentences.

Little trip.
To go or going

14.1 The to-form and the ing-form

For the verb 'go', we call the form 'to go' the 'to-form' (or 'infinitive').

Everybody remembers Star Trek: "To boldly go where no man has gone before."

The 'ing-form': 'going', is the matching pair for the 'to-form' (grammarians call the 'ing-form': 'gerund').

Examples:

To solve the quadratic equation, we first need to find the values of a, b, and c. (to-form)

———

Simplifying algebraic expressions is crucial for solving equations efficiently. Graphing the function allows us to visualize its behavior. (ing-form)

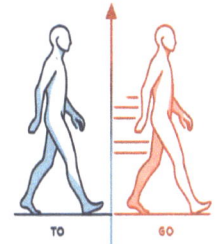

Figure 14.1: To go and going (Vecteezy Library)

———

The goal is to simplify the expression before substituting the values of the variables. (to-form)

Supplementary Information The online version contains supplementary material available at https://doi.org/10.1007/978-3-031-65161-8_14.

Calculating the determinant of a matrix is a step in finding its inverse. (ing-form)

Interchanging the to-form with the ing-form can lead to confusion:

I stopped to smoke.

vs.

I stopped smoking.

14.2   Full and incomplete infinitives

Since we will discuss the 'to-form' without the 'to', it is time to use the term 'infinitive'. An 'infinitive' is the basic form of a verb, preceded by the word 'to'. But it can appear in two forms: as 'full infinitive' when 'to' is included, and as 'incomplete infinitive' when 'to' is omitted.

14.2.1   Full infinitive

The full infinitive is used in various ways, including as a noun, an adjective, or an adverb to explain the reason or objective behind an action.

Typical examples:

To learn is important.

—————

I want to prove this theorem.

—————

She seems to understand the problem.

—————

To prove this theorem, we first assume…

—————

To integrate this function is to find the area under its curve.

—————

The method is easy to apply.

14.2.2   Incomplete infinitive

Incomplete infinitives often appear in math writing, especially after helping verbs (can, could, will, would, shall, should, may, might, must) or certain other verbs (let, make, see, hear, help, and sometimes feel, watch, notice).

Examples:

We can prove the theorem using…

—————

We will return to this topic in Topic 6.

—————

This lemma lets us simplify the equation.

———

We see the function converge rapidly.

14.3   Split infinitive

A 'split infinitive' occurs when an adverb or other word is inserted between 'to' and the verb in an infinitive as in the 'Star Trek' quote.

Borrowing a rule of Latin grammar, grammarians of English traditionally considered split infinitives incorrect. In Latin, infinitives are single words and thus incapable of being split. However, this view has changed over time. Modern English usage accepts split infinitives, especially when they enhance clarity or produce a more natural-sounding sentence.

Examples:

To quickly develop into a new area of research requires courage.

———

To clearly delineate the difference…

———

To simply define trunc T, we start with the string S.

Figure 14.2:   Split infinitive
(Vecteezy Library)

Little dangling

This is not just one more topic. It is very important. The problems it discusses are widespread, and not so easy to find. I confess, most of the time, it really sounds OK to me.

To 'dangle' means to hang or swing loosely, especially with an end that moves freely. In grammar, when we talk about 'dangling' modifiers, it describes a word or phrase that is improperly linked to the words it's supposed to modify.

In the previous topic, we discussed the '-ing form' of a verb. Similarly, there is an '-ed form': we add -'ed' to the base form of a verb. So 'support' becomes 'supported' and 'form' becomes 'formed'. Irregular verbs are exceptions: 'brake' becomes 'broken' and 'write' becomes 'written'.

Figure 15.1: Dangle
(Vecteezy Library)

A 'participle' is the '-ing form' or the '-ed form' of a verb. This term now allow us to introduce 'dangling participles', the subject of this topic. We make the introduction with an example:

Determined to solve the equation, the solution seemed elusive. 😖

Here, "Determined to solve the equation" is intended to modify the mathemati-

Supplementary Information The online version contains supplementary material available at https://doi.org/10.1007/978-3-031-65161-8_15.

cian attempting the task. Instead, it modifies "the solution", suggesting that the solution itself is making an effort to solve the equation. LOL

Dangling participle

A 'dangling participle' is a serious grammatical error that occurs when a participial phrase (an adjective phrase that starts with a participle) is not clearly or logically related to the noun it is intended to modify. This usually happens because the noun being modified is missing from the sentence, leaving the participle 'dangling', with nothing to modify.

Here is another example:

Running to catch the bus, my book fell out of my bag. 😫

In this sentence, "Running to catch the bus" is a participial phrase meant to describe the action of the subject. However, the subject immediately following this phrase is "my book", which cannot run. The intended meaning is likely that the speaker or another person was running to catch the bus when the book fell out of their bag. Correct it:

Running to catch the bus, I dropped my book. 🙂

Dangling participles disrupt the clarity of writing and can lead to misinterpretation or unintentional humor. They're often easily fixed by restructuring the sentence so that the participle clearly modifies the intended noun or by introducing the intended subject immediately after the participial phrase.

Using the quadratic formula, the roots' nature was immediately clear. 😫

This sentence leaves unclear who is using the quadratic formula.

When we used the quadratic formula, the nature of the roots became immediately clear. 🙂

Math examples

Having calculated the final variable, the theorem was proven conclusively. 😫
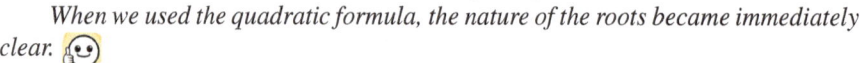
After we had calculated the final variable, we conclusively proved the theorem. 🙂

"Having calculated the final variable" seems to modify "the theorem"— a theorem cannot calculate. The corrected sentence clarifies that "we" did the calculating and proving.

———

Working through the complex proof, frustration became a constant companion. 😫

As we worked through the complex proof, frustration became our constant companion. 🙂

———

Upon reviewing the equations, mistakes were found in the calculations. 😫

Upon reviewing the equations, we found mistakes in the calculations. 🙂

———

Using a new algorithm, the problem's complexity was significantly reduced.
By using a new algorithm, we significantly reduced the problem's complexity.

Strategies for avoidance and correction

Identify the action Begin by identifying the action your participle describes. Ensure this action is clearly and logically connected to a doer mentioned in the sentence.

Rephrase for clarity Often, the simplest solution to a dangling participle is to rephrase the sentence to include the actor explicitly. For example, "After calculating the integral, the theorem's proof was complete" could be revised to "After we calculated the integral, we completed the theorem's proof".

Restructure with active voice Many dangling participles result from attempts to use a passive construction where an active voice would be more direct and clear. Consider using an active voice to directly connect the action with its doer.

TOPIC

16

Little numbers. Less and fewer, both and two

16.1 Less and fewer

'Mark Twain' is a term used by the pilots of the Mississippi river to indicate a depth of two fathoms (or twelve feet), which is considered safe for steamboats.

Since the author's name Mark Twain has the number two in it, we start illustrating the use of 'Less and fewer' with a quote:

Mark Twain: "If we had less statesmanship, we could get along with fewer battleships."[1]

The distinction between 'less and fewer' is straightforward: 'fewer' refers to items that can be counted (discrete quantities), while 'less' is used for measurable (continuous) quantities. (Halmos agrees, see Appendix C.) Let's explore examples to clarify their proper uses and illustrate common

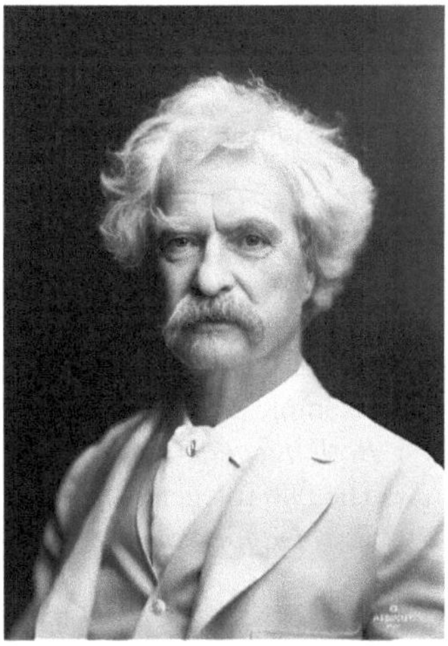

Figure 16.1: Mark Twain

[1]Wisdom Trove, `https:wisdomtrove.com`

Supplementary Information The online version contains supplementary material available at https://doi.org/10.1007/978-3-031-65161-8_16.

G. Grätzer, *The Little Book of Math into English*, https://doi.org/10.1007/978-3-031-65161-8_16

misuses.

Examples of correct usage

There are fewer students in class today than yesterday.
Students can be counted.

————

I solved fewer math problems today than I did last week.
Problems are discrete items.

————

I spent less time on homework today than I did last week.
Time is a continuous quantity.

Examples of incorrect usage

This recipe requires fewer sugar than the other one.
Sugar, when not specified in units like cups or tablespoons, should be treated as a continuous quantity.

————

There are less cars in the parking lot today.
There are fewer cars in the parking lot today.

————

There are less people in line today than yesterday.
There are fewer people in line today than yesterday.

This is an easy topic made more complicated by misuse in the media:

Does Israel's Post War Plan for Gaza Include Less Gazans?

Dutch universities plan to limit the amount of international students.

16.2   Both and two

The definition is straightforward: 'both' refers to a shared property of two items, while 'two' counts the items.

Examples:

Both integers a and b are even.
The two integers share the property of being even.

To solve for x in the equation ax + b = 0, both a and b must be real numbers.

It correctly specifies a condition involving two elements.

————

For a point to lie on the line $y = mx + c$, it must satisfy both the equation of the line and any constraints given by the problem.
This accurately describes that the two conditions must be met concurrently.

Figure 16.2: Both and two (Vecteezy Library)

Both of the two solutions to the quadratic equation are real.
The phrase 'both of the two' is redundant because 'both' already implies 'two'.

Two of the roots of the polynomial are complex conjugates.
This specifies a quantity related to the elements described.

A linear equation in two variables has an infinite number of solutions.
It specifies the number of variables involved.

16.3 *'The two'*

The topic of the previous section is so easy to understand. And then you come across
a sentence like this:

The two approaches to the problem yielded different results.

This is certainly well phrased, but does not quite agree with the definitions in
the previous section. Here, 'the two' almost sounds like 'both'.

In general, 'the two' focuses on the distinct aspects of two items. The word 'both'
emphasizes their shared attributes, two items together, not necessarily distinguishing
between them. The distinction comes down to nuance rather than a strict difference
in meaning.

Examples:

Let x_1 and x_2 be the two solutions of the quadratic equation $ax^2 + bx^2 + c = 0$.

This emphasizes that there are two solutions.

Both solutions of the quadratic equation satisfy the condition $x^2 - 4 = 0$.
So each solution satisfies the condition.

The sets $A = \{1, 2\}$ and $B = \{3, 4\}$ have distinct elements.

This highlights the individuality of each set. 'Both' sets have exactly two elements, focusing on the common characteristic shared by the sets.

———

The two sets $A = \{a\}$ and $B = \{b\}$ are singletons.

Focusing on the common characteristic shared by the sets.

16.4   Word repetition

Figure 16.3: The chaos of word repetition (Vecteezy Library)

16.4.1   General examples

I suggest keeping an eye on word repetition in the final editing.

Use synonyms

John, the manager, discussed the budget issues. Later, the manager offered solutions to the budget problems. 😩

John, the manager, discussed the budget issues. Later, he offered solutions to the financial problems. 😌

Vary sentence structure

The teacher explained the lesson. The teacher then gave the students homework. 😩

After explaining the lesson, the teacher assigned homework to the students. 😌

Combine sentences

John went to the store. John bought milk.

    John went to the store and bought milk.

Use pronouns to replace nouns

Sarah said that Sarah would attend the meeting if Sarah finished her work early.

    Sarah said she would attend the meeting if she finished her work early.

Eliminate unnecessary words

The results of the survey clearly showed that the majority of participants prefer to have more options.

    The survey results clearly showed that the majority of participants prefer more options.

Change passive voice to active voice

The book was read by the students before the class.

    The students read the book before the class.

Use specific details over general terms

The dog played with its toy. The dog seemed happy with the toy.

    The dog wagged its tail vigorously while playing with the squeaky ball.

Figure 16.4: Math examples (Vecteezy Library)

16.4.2   Math examples

Describing processes

To solve the equation, add 5 to both sides of the equation. After adding 5, divide both sides of the equation by 3.

To solve the equation, first add 5 to both sides, then divide the results by 3. 🙃

Explaining solutions

The area of the rectangle is found by multiplying the length by the width. The length is 5 cm and the width is 3 cm. So, the area of the rectangle is 15 cm².

The area of a rectangle, calculated as length times width, is 15 cm² for dimensions of 5 cm by 3 cm. 🙃

Writing proofs

Assume that a and b are even numbers. Since a and b are even, $a = 2k$ and $b = 2m$, where k and m are integers. The sum $a + b$ is then $2k + 2m = 2(k + m)$, so it is even.

Let $a = 2k$ and $b = 2m$ represent even numbers, where k and m are integers. Then the sum $a + b = 2(k + m)$ is also even. 🙃

Demonstrating calculations

To find the derivative of $f(x) = x^2$, use the power rule. Apply the power rule to get $2x$.

To find the derivative of $f(x) = x^2$, apply the power rule to obtain $2x$. 🙃

Presenting results

The experiment showed that the reaction time decreases as the temperature increases. When the temperature was increased, the reaction time was shorter.

The experiment revealed a decrease in reaction time with increasing temperature, demonstrating a direct correlation between temperature and the speed of the reaction.

Explaining concepts

A prime number is a number that has exactly two distinct positive divisors: 1 and itself. For a number to be prime, it must be divisible only by 1 and itself.

By definition, a prime number has exactly two distinct positive divisors, namely, 1 and itself. 🙃

16.5   Parallel structure

Figure 16.5: Parallel structures (Vecteezy Library)

Parallel structure is a grammatical and stylistic device in English: using the same pattern of words to show that two or more ideas have the same level of importance. Parallel structure helps to create readability.

Parallel words:

She likes hiking, biking, and swimming. 😊

Parallel phrases:

He is responsible for cooking dinner, cleaning the house, and taking care of the dog. 😊

Parallel clauses:

The teacher said that we should listen carefully, that we should take detailed notes, and that we should ask questions if we don't understand. 😊

Math examples

A square is a rectangle with all sides equal, and a rhombus is a parallelogram with all sides equal. 😊

_____

The sum of the angles in a triangle is 180 degrees. The sum of the angles in a quadrilateral is 360 degrees. 😊

_____

First, we demonstrate the base case. Second, we assume the statement is true for k and prove the statement for $k + 1$.

In a field, addition is associative, multiplication is commutative, and multiplication distributes over addition.

The area of a circle is πr^2, while the circumference is $2\pi r$.

16.5.1   Beware of the pitfalls

When discussing the importance of maintaining consistency in math writing, it's crucial to avoid common pitfalls that can make the content harder to understand. Let's look at examples.

Mismatched elements. Combining different grammatical structures within a list, can confuse the reader and obscure the meaning.

> *To solve the equation, you must first identify the variables, solving the quadratic formula, and understanding the graph's shape.*

In this sentence, there's a mixture of verbs ("identify") and gerund phrases ("solving the quadratic formula", "understanding the graph's shape"). This mismatch creates a jarring list that's hard to follow.

> *To solve the equation, you must first identify the variables, solve the quadratic formula, and understand the graph's shape.*

Isn't this sentence so much easier to read?

Shift tenses within parallel structures

> *When calculating the area of a circle, you first determine the radius and then multiplied it by itself before applying π.*

The shift from present tense ("determine") to past tense "multiplied") within a procedural explanation disrupts the parallel structure necessary for clarity.

> *When calculating the area of a circle, you first determine the radius and then multiply it by itself before applying π.*

Both active and passive voices in parallel elements

> *To find the limit of a function as x approaches infinity, you must compare the rates of growth in the numerator and the denominator, and it is determined by applying L'Hôpital's Rule if necessary.*

The sentence starts with an active voice ("you must compare") but switches to passive voice ("it is determined"), which can confuse the reader regarding the subject performing the actions. Also, the pronoun 'it' has ambiguous referent.

> *To find the limit of a function as x approaches infinity, you must compare the rates of growth in the numerator and denominator, and then apply L'Hôpital's Rule if necessary.*

Little pause.
Comma
before if

17.1   Dependent and independent clauses

To correctly understand and apply the use of commas in sentences involving dependent and independent clauses, especially with conditional sentences and conjunctions, let's clarify these concepts with examples and explanations.

Understanding dependent and independent clauses

Figure 17.1: Dependent Klaus (Vecteezy Library)

An 'independent clause' is a group of words that contains a subject and verb and expresses a complete thought. It can stand alone as a sentence. A 'dependent clause' cannot stand alone as a sentence because it does not express a complete thought. It usually provides additional information to an independent clause and is often introduced by conjunctions like 'if', 'because', 'although', and so forth.

Supplementary Information The online version contains supplementary material available at https://doi.org/10.1007/978-3-031-65161-8_17.

Commas with conditional sentences

Conditional sentences often use 'if' to introduce a condition. The use of commas depends on the order of the clauses.

When the dependent clause (condition) comes first, a comma is used to separate it from the independent clause:

Example:

If L is distributive, then K is flat. 😊

Here, "If L is distributive" is the dependent clause that sets a condition for "K is flat", the independent clause.

When the independent clause comes first, no comma is needed before the dependent clause:

Example:

K is flat if L is distributive. 😊

In this structure, the independent clause "K is flat" is followed by the dependent clause "if L is distributive", so no comma is used.

Using commas in complex sentences

A 'complex sentence' consists of one independent clause and at least one dependent clause. The placement of the comma depends on the order of these clauses.

Dependent clause followed by an independent clause: Use a comma after the dependent clause.

Examples:

Because (5) holds, the lattice L is regular. 😊

Independent clause followed by a dependent clause: Do not use a comma before the dependent clause.

The lattice L is regular because (5) holds. 😊

———

If L is distributive, then K is flat. 😊

Comma is used because the dependent clause comes first.

The lattice L is regular if (5) holds. 😊

No comma is needed because the independent clause comes first.

In both examples, the clause following 'if' is a dependent clause because it provides a condition and does not stand alone as a complete thought. Thus, when it precedes the main (independent) clause, a comma is used. When it follows the main clause, no comma is necessary.

In summary, the key to using commas correctly with dependent and independent clauses, especially in conditional sentences, is understanding the order of the clauses and whether the clause introduces additional information or sets a condition. Remember, a comma is typically used after a dependent clause when it precedes the independent clause in a sentence. When a dependent clause follows the main clause, a comma is generally not required.

Figure 17.2: Complex sentence (Vecteezy Library)

Little ambiguity

18.1 Ambiguity

Mathematicians follow Humpty Dumpty's dictum: "When I use a word, it means just what I choose it to mean" (Lewis Carroll [4]). They use everday words to name abstract constructs. Ambiguity follows.

18.1.1 Examples

Ambiguous words and phrases can lead to misunderstandings. These are words or phrases with more than one interpretation.

Figure 18.1: Ambiguous (Vecteezy Library)

Supplementary Information The online version contains supplementary material available at https://doi.org/10.1007/978-3-031-65161-8_18.

Examples:

Normal

The word 'normal' means typical or standard. However, in math, 'normal' is an overused—but nonetheless often standard—adjective: normal distribution in statistics, a normal subgroup in algebra, or a normal line in geometry.

    A normal family of complex functions.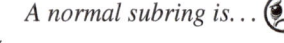
is confusing.
    Instead of
    A normal subring is... 😣
try
    A standard subring is... 🙂
    'Standard' is a nice synonym of 'normal', not so overused.

Ring

Eng *A ring is a piece of jewelry.* 🙂

———

    Math *However, a ring is also an algebraic structure.* 🙂
    "He studied the ring" could lead to confusion if the audience does not know whether the subject is a mathematician or a jeweler.

Function

Everyday *The term 'function' refers to the purpose that something is designed to do.* 🙂

———

    Math *A 'function' is a relation between a set of inputs and a set of permissible outputs with the property that each input is related to exactly one output.* 🙂
    "The function of the algorithm" could be ambiguous if it's unclear whether the discussion is about the purpose of the algorithm or a math function within the algorithm.

Series

Everyday The word 'series' refers to a sequence of events, stories, or television episodes.
    Math A 'series' is the sequence of partial sums of a given sequence.
    After completing the series, we left town. 😣
may refer to watching a set of lectures or solving a series of exercises.

Field

Everyday A 'field' might denote an area of open land.
    Math A 'field' is an algebraic structure.

Figure 18.2: Bull out standing in the field (Vecteezy Library)

"His work in the field" is ambiguous; clarify whether it refers to agricultural work, fieldwork in a scientific study, or research in a specific area of math.

Integral

Everyday 'Integral' means essential or necessary for completeness.

Math 'Integral' is a fundamental concept in calculus and analysis.

Finding the area under a curve requires an integral approach.
It's unclear whether this sentence refers to an essential approach or the math process of integration.

Rational

Everyday 'Rational' means based on or in accordance with reason or logic.

Math A number is 'rational' if it can be expressed as the quotient of two integers.

The discussion on rational decisions led to a surprising conclusion.
This could be misinterpreted as a discussion about decisions based on rational numbers.

Complex

Everyday 'Complex': consisting of many different and connected parts; not easy to analyze or understand.

Math A number is 'complex' if it can be written in the form $a + bi$, where a and b are real numbers and i is the square root of -1.

The complex problem was finally solved.
Was the problem intricate or related to complex numbers?

Mean

Everyday 'Mean' is the average or middle point between extremes.
Math 'Mean' refers to arithmetic mean (average), geometric mean, or harmonic mean.

   The mean temperature for July shows a significant deviation from the norm. 😞
It is unclear whether this refers to an average temperature or 'mean' has another meaning.

Radical

Often, the ambiguity of a word is not binary: general English use and math use. As an example, consider the word 'radical', it has many meanings across different contexts.

**FREE RADICAL
AND NORMAL MOLECULE**

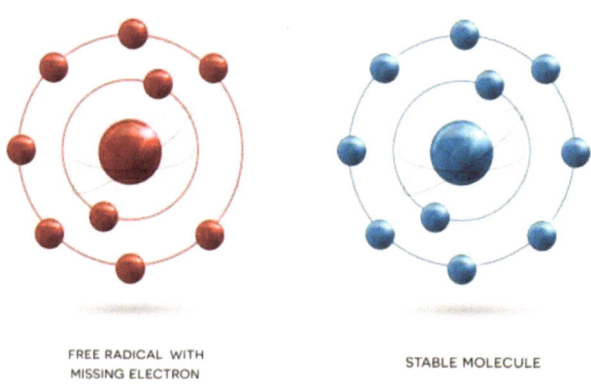

FREE RADICAL WITH
MISSING ELECTRON

STABLE MOLECULE

Figure 18.3: Radicals (Vecteezy Library)

General use

In a broad sense, 'radical' refers to anything that represents a significant departure from the norm, tradition, or conventional expectations. It implies a fundamental or profound change.

   Implementing a four-day workweek would be a radical departure in the corporate world. 😊

Math

In math, 'radical' refers to the root of a number, especially the square root.

The radical symbol is used to denote the square root of a number, so $\sqrt{16} = 4$.

Or 'radical' refers to the construct for associative algebras or rings.

Politics and Social Science

In this context, 'radical' describes individuals, groups, ideologies, or movements that advocate for significant, fundamental reforms or revolutionary changes in society.

The suffragette movement in the early 20th century was considered radical for its time because it sought to overturn the conventional norms by fighting for women's right to vote. 😊

Chemistry

In chemistry, a 'radical' is an atom, molecule, or ion with unpaired valence electrons, making it highly reactive.

The hydroxyl radical is a radical involved in many important chemical reactions in the atmosphere, including the decomposition of pollutants. 😊

Language and Linguistics

In the study of languages, especially with regard to the classification of Chinese characters, a 'radical' is a component of a character often used to indicate some aspect of its meaning or pronunciation.

For instance, the character for 'river' contains the 'water' radical, which suggests a relationship to water.

18.1.2 Homonyms

Words such as 'record' are called 'homonyms': a type of ambiguity where two words sound the same or are spelled the same but have different meanings.

Etymology studies the history and origins of words.

For example, 'record' as a noun comes originally from the Latin 'recordari', which com-bines 're-' (again) with 'cor' (heart), suggesting 'to remember by heart'.

Figure 18.4: Entomology (Vecteezy Library)

As a verb, 'record' shares the same roots but diverged in usage to specifically refer to the action of capturing or registering information.

18.2   More examples

Invalid

'Invalid' (second syllable stressed, in-VAL-id). Not valid or logically sound in the context of math arguments or proofs.

'Invalid' (first syllable stressed, IN-va-lid). A person made weak or disabled by illness or injury.

The proof was considered invalid because of a technical error.

_____

After the accident, he became an invalid and could no longer attend the math conference.

Minute

'Minute' (first syllable stressed, MIN-ute). A unit of time equal to 60 seconds, often used in the timing of math problems or experiments.

'Minute' (second syllable stressed, mi-NYOOT). Extremely small, as in a minute difference or error in a math calculation or measurement.

The experiment required minute measurements of time, accurate to the millisecond.

_____

The difference in results was so minute that it was almost negligible, but it proved to be significant in the complex math model.

Refuse

'Refuse' (first syllable stressed, REF-us). Waste material, not directly related to math but can be involved in statistical studies about waste management.

'Refuse' (second syllable stressed, re-FUSE). To decline to accept or allow, which could be used in the context of disproving a math theory or hypothesis.

'Refuse' (first syllable stressed, REF-us). In statistics:

The study on household waste produced a large set of refuse data for statistical analysis.

'Refuse' (second syllable stressed re-FUSE). To reject a hypothesis:

The mathematician refused to accept the hypothesis without further proof, demonstrating a critical approach to the problem.

Project

'Project' (first syllable stressed, PROJ-ect). A planned endeavor, potentially involving math research or study.

'Project' (second syllable stressed, pro-JECT). To cast forward or predict, often involving math models or forecasts.

'Project' (first syllable stressed) in math:

She led a research project aiming to develop new algorithms for machine learning. 😊

'Project' (second syllable stressed) future trends:

Using statistical models, we can project future trends in climate change with a reasonable degree of accuracy. 😊

Attribute

'Attribute' (first syllable stressed, ATTR-i-bute), a noun. Regard as a characteristic or inherent part; for instance, in descriptive statistics or data analysis.

'Attribute' (second syllable stressed, at-TRIB-ute), a verb. Regard as caused by a certain factor, used in statistical analysis or math modeling.

One important attribute of the dataset is its high level of accuracy, which is crucial for our analysis. 😊

'Attribute' (second syllable stressed, at-TRIB-ute) causality:

The researchers were careful not to attribute the changes in temperature solely to increased carbon emissions without considering other factors. 😊

Do not misspell

Misspellings are a good source of ambiguity: 'to' vs. 'too' vs. 'two'.

My advice: Do not misspell. LOL. For an example, see Section B.6.1.

Little active vs. passive

19.1 The problem

Mission Impossible: 'active sentences' vs. 'passive sentences'.

Should we write the active sentence:

Bjarni discovered Tre-ideals in 2021. 😊

or the passive sentence:

Tre-ideals were discovered by Bjarni in 2021. 😵

It seems like a simple question with an obvious answer. The first option is shorter and more straightforward, while the second is lengthier and more complex.

Figure 19.1: Bjarni

While discussing this question with a well-established mathematician, he confessed to struggling with this.

Supplementary Information The online version contains supplementary material available at https://doi.org/10.1007/978-3-031-65161-8_19.

Having read and written math for decades, he so often encountered passive sentences that they seemed natural. To him, a passive sentence feels natural.

When I submitted my first math book to the publisher, the editor told me that she had begun changing my passive sentences to active ones, but after realizing that I used passive structure consistently, she stopped changing them.

My 'Mission Impossible' is an attempt to convince you that, as a rule, active sentences are preferable. LOL. 😜

19.2   Some examples

Figure 19.2: Mark the letter A with a bar (Vecteezy Library)

Here is a typical example of a passive sentence:

It has been conjectured that any algebraic distributive lattice is the congruence lattice of some lattice, but this has not been proven in full generality. 🙁

I rewrite it as an active sentence:

We do not know whether the conjecture is true that every algebraic that every algebraic distributive lattice is the congruence lattice of some lattice. 🙂

Jerzy Trzeciak [20] categorizes the passive sentences used in math into five types. (He does not state a position on whether we should use passive sentences.) We will now present Trzeciak's examples of these five types.

I recommend comparing your passive sentences to the samples for these types and converting them as the samples indicate, to make them direct.

Type 1: "we do something"

> Passive:
> *This identity is established by connecting (8) and (9).*
> Active:
> *We establish this identity by connecting equations (8) and (9).*

> Passive:
> *This difficulty is avoided in the above paragraph.*
> Active: *We avoid this difficulty in the paragraph above.*

> Passive: *When x is substituted into (3), an analogous description of K is obtained.*
>
> Active: *We obtain an analogous description of K by substituting x into the equation (3).*

> Passive: *Nothing is assumed concerning the expectation of X.*
> Active: *We do not assume anything concerning the expectation of X.*

Type 2: "we prove that X is"

> Passive:
> *The function f is easily shown to have property (P).*
> Active:
> *It is easy to show that the function f has property (P).*

> Passive:
> *A space T is is said to be regular if property (R) holds.*
> Active:
> *We say that a space T is regular if property (R) holds.*

> Passive: *This equation is known to hold for $x > 2$.*
> Active: *We know that this equation holds for $x > 2$.*

Type 3

Replacing the phrase "we give an object X a structure Y":

Passive:

Note that E can be given a complex structure by (9).

Active:

Note that (9) gives E a complex structure. 🙂

———

Passive:

The letter A is marked by a bar to indicate that (5) holds. ☹

Active:

We mark the letter A by a bar to indicate that (5) holds. 🙂

Type 4

Replacing the phrase "we act on something":

Passive:

This order behaves well when g is acted upon by an operator. ☹

Active:

This order behaves well when an operator acts on g.

———

Passive:

So all the terms of (5) are accounted for. ☹

Active:

So this accounts for all the terms of (5).

———

Passive:

The preceding observation, when looked at from a more general point of view leads to the conclusion. ☹

Active:

Looking at the preceding observation from a more general point of view leads to the conclusion. 🙂

Type 5

Meaning "which will be (proved, etc.)":

Passive:

Before stating the result to be proved, we provide an example. ☹

Active:

Before stating the result to prove, we provide an example.

———

Passive:

This is a special case of convolutions to be introduced in Topic 8.

Active:

In Topic 8, we will introduce this special case of convolutions.

———

Passive:

We conclude with two simple lemmas to be used in the next sec-tion.

Active:

We conclude with two simple lemmas for use in the next section.

19.3   Please

Please, consider using active sentences as much as possible. This, by itself, will make a sizeable contribution to the 80% we set out to achieve in the Introduction.

19.4   Little direct vs. indirect

Figure 19.3: Direct vs.indirect (Vecteezy Library)

Direct vs. indirect is similar to active vs. passive.

But while active vs. passive is of crucial importance in math, direct vs. indirect is not—we seldom quote speech.

And note that the transformation from indirect to direct needs special care because of changes in tense, pronouns, and time expressions.
Direct:

He said, "I am eating."

Indirect:

He said that he was eating.

_____

Direct:

Mary said, "I will help you tomorrow".

Indirect:

Mary said that she would help me the next day.

_____

Direct:

He said, "I will see you today."

Indirect:

He said that he would see me today.

_____

Direct:

She said, "I have finished my work today."

Indirect:

She said that she had finished her work today.

_____

Direct:

He asked, "Are you coming?"

Indirect:

He asked if I was coming.

or

Indirect:

He asked if I were coming.

Little speed. Faster and fastests

To compare two or more items, we use the 'comparative form': 'faster'; to describe an item at the extreme upper end of a spectrum, we use the 'superlative mode': 'fastest'.

   Make sure they are properly used.

Figure 20.1: Faster and fastests (Vecteezy Library)

Supplementary Information The online version contains supplementary material available at https://doi.org/10.1007/978-3-031-65161-8_20.

G. Grätzer, *The Little Book of Math into English*, https://doi.org/10.1007/978-3-031-65161-8_20

Of all the algorithms tested, Algorithm A was faster. 😖

Of all the algorithms tested, Algorithm A was the fastest. 🙂

When comparing the two datasets, Dataset 1 was the most variable. 😖

When we compare the two datasets, Dataset 1 was more variable than Dataset 2. 😖

This would be great except that it's 'dangling' (see Topic 15). Lets 'undangle' it:

When we compared the two datasets, Dataset 1 was more variable than Dataset 2. 🙂

In our comparison of the two datasets, Dataset 1 was found to be more variable than Dataset 2. 🙂

Dataset 1 was more variable than Dataset 2 upon comparing them. 🙂

Little distinctions

21.1 'a' vs. 'the' vs. blank

If you are a mathematician whose first language is English, you **should skip** this section; I would not want to risk making you feel superior.—For mathematicians whose first language is not English, the nuances of article usage in English (specifically 'a', 'the', and no article (blank)) can pose significant challenges. This often represents a common stumbling block for Russian (and other Slavic) speakers. We'll discuss this topic in detail.

Figure 21.1: 'a' and 'the'
(Vecteezy Library)

Supplementary Information The online version contains supplementary material available at https://doi.org/10.1007/978-3-031-65161-8_21.

21.2   *Basics*

The indefinite article: 'a'

The indefinite article 'a' ('an' before a word beginning with a vowel sound) is used in English before singular, countable nouns which are mentioned for the first time or are not specific to the listener or reader. In math writing, 'a' is often used when introducing a new variable, concept, or entity whose specifics are not yet defined.

Let a be a prime number.

The definite article: 'the'

The definite article 'the' is used before both singular and plural nouns that are specific and known to the reader, usually previously mentioned or universally acknowledged within the text.

Once the equation is established, the solution becomes evident.

No article

No article is used with plural or uncountable nouns when speaking about them in a general sense. This rule also applies to abstract concepts and most names of disciplines or sciences.

Quadratic equations are integral to algebraic studies. 🙂

21.3   *Typical mistakes*

Russian speakers might find the English system of articles challenging due to the absence of definite and indefinite articles in the Russian language. Here are some common pitfalls and how to avoid them:

Overgeneralization

Solution to equation is found easily. 🙁
   The solution to the equation is found easily. 🙂
   'The' specifies the particular solution to a previously defined or understood equation.

Unnecessary articles

The math is a beautiful field. 🙁
   Mathematics is a beautiful field. 🙂

Avoid articles with uncountable nouns used in a general sense.

Rules specific to math writing

The language of math has its own set of conventions regarding articles.

Introducing new terms
Let n be an integer. ☺

Use 'a' ('an') when a term is introduced as it represents one of many possible values or instances.

Referring to defined terms
The matrix A is invertible. ☺

Use 'the' when referring back to specific items or concepts that have been previously defined or are unique within the context.

General statements and theorems Avoid articles when making general statements about subjects or concepts.

A zero is a placeholder in positional number systems. ☹

Zero is a placeholder in positional number systems. ☺

———

A Pythagorean theorem is useful for calculating distances. ☹

The Pythagorean theorem is useful for calculating distances. ☺

Figure 21.2: Russian A (Vecteezy Library)

21.4   More examples

Using 'a' (indefinite article)

Introducing a new element.

———

Let v be vector in space.😩

Let v be a vector in the space. 😗

'a' is used because the vector *v* is being introduced for the first time and is not yet specific.

———

We need eigenvalue for this calculation.😩

We need an eigenvalue for this calculation. 😊

Use 'an' before vowel sounds; an eigenvalue is mentioned as a generic example of eigenvalues.

Using 'the' (definite article)

Referring to specific entities known to the reader

Integral of function gives area under curve.😩

The integral of the function gives the area under the curve. 😊

'the' specifies the particular integral and function discussed, and the curve related to them.

When detailing steps or continuing an explanation

Substitute value into equation to find solution.😩

Substitute the value into the equation to find the solution. 😊

'the' is used to refer to specific value and equation already discussed or understood from context.

No article

The quadratic functions are useful in engineering.😩

Quadratic functions are useful in engineering. 😊

No article is used when referring to things in general; quadratic functions here refer to all such functions, not a specific set.

Fields of study

The algebra is a branch of math that studies structures.

Algebra is a branch of math that studies structures. 😊

Fields of study do not generally take an article unless specified in a particular context.

Theorems and principles in general

The Pythagoras' theorem is fundamental in geometry. 😩

Pythagoras' theorem is fundamental in geometry. 😊

The name of a theorem does not require an article unless talking about a specific instance or application of it.

21.5 Maybe vs. may be

The distinction between 'maybe' and 'may be' is important.

'Maybe' is an adverb used to indicate uncertainty or possibility. It typically precedes the subject or is used in response to a question.

Maybe we can solve this equation using a different method. 😊

―――――

Can we prove this? Maybe. 😊

'May be' is a verb phrase consisting of the helping verb (see Section 14.2.2) 'may' and the verb 'be'. It expresses a possibility concerning a state or condition.

The solution may be more complex than we initially thought. 😊

―――――

There may be a mistake in the proof of this theorem.

'Maybe' often comes at the beginning of a sentence or clause, or as a standalone response.

'May be' is part of the predicate and follows the subject.

Figure 21.3: Maybe and may be (Vecteezy Library)

Consider the sentence structure and what you wish to express. If you're indicating uncertainty about an entire statement, 'maybe' is appropriate. If you're discussing the potential characteristics or state of a specific subject, 'may be' is the correct choice.

More examples:

Maybe we should consider a geometric interpretation of the problem to gain new insights. 😊

———

Maybe next Tuesday's seminar could focus on recent advancements in topology. 😊

———

Is this algorithm efficient for all cases? Maybe, but further testing is needed to confirm that this is so. 😊

———

The discrepancy in the data may be due to measurement error. 😊

———

There may be a correlation between these two variables that we haven't yet explored. 😊

———

The observed phenomenon may be an instance of the so-called butterfly effect in chaotic systems. 😊

"May be" Examples:

"She may be at the library now." "There may be a meeting tomorrow." "Might be"

Might be vs. may be

Both 'might be' and 'may be' convey possibility; 'may be' often suggests a higher likelihood and is slightly more formal, while, 'might be' suggests a lower likelihood and is more casual.

Examples:

He might be running late. 😊

There might be a thunderstorm tonight. 😊Summary

21.6   Further vs. farther

The distinction between 'further' and 'farther' is also subtle. Both are appropriate for distance or extent, but they are used in slightly different contexts.

Figure 21.4: Further and farther (Vecteezy Library)

'Farther' refers to physical distance. It is the comparative form of 'far' when discussing tangible, measurable distances.

To prove the theorem, we need to extend the line segment farther beyond the initial point. 😊

———

In constructing the geometric proof, the distance from point A to point B needs to be farther to maintain the properties of the triangle. 😊

'Further' refers to a metaphorical or figurative distance, or to indicate a greater degree or extent in non-physical contexts. It can also imply addition or advancement in time or actions.

More examples:

We need to discuss this matter further. 😊

———

Further exploration of algebraic structures may lead to a deeper understanding of symmetry in mathematics. 😊

———

Further research is required to understand the implications of this hypothesis. 😊

The distinction between 'farther' and 'further' is becoming increasingly blurred in modern usage, and in many cases, they are used interchangeably.

Note that there is the word 'furthermore', an adverb used to introduce a point that adds to or supports what has already been said. It's similar to 'in addition', 'moreover', or 'besides'.

The plan is not feasible at this time. Furthermore, we do not have the necessary funds to support it. 😊

'Farthermore'? There is no such word.

21.7   Shall vs. will

Both 'shall' and 'will' are helping verbs indicating future tense.

The 80% Rule. Always use 'will'.

But do not tell this to General Douglas MacArthur. The title of his famous speech on March 20, 1942:

I Shall Return.

Shall

'Shall' is traditionally used with 'I 'and 'we' to form the future tense. It implies formality, determination, or obligation. In legal documents, it is used to state a legal requirement or obligation.

We shall now prove the following theorem to establish the foundation for our argument. 😊

We now present a proof, but we shall do better in Chapter 6 with a really short and clever proof. 😊

Will

'Will' is used with all subjects (I, you, he, she, it, we, they) to indicate the future tense. It expresses a future action or a willingness or promise to perform a future action.

Example:

We will introduce a new method to solve nonlinear equations more efficiently. 😊

More illustrations for 'shall'

In the next section, we shall explore the implications of these findings on the theory of probability. 😊

We shall prove the existence of an infinite number of prime numbers using a methodological approach outlined in Chapter 4. 😊

Participants shall submit their solutions to the problem set by the deadline indicated in the syllabus. 😊

———

We shall address all potential counterexamples to ensure the robustness of our conjecture. 😊

More illustrations for 'will'

This discovery will likely open new avenues for research in computational algebra. 😊

———

We will start by examining the dataset for any anomalies before proceeding with the analysis. 😊

———

The research team will[1] provide updates on the progress of the project at the weekly seminar. 😊

———

I will review the literature to identify any similar approaches that have been previously attempted. 😊

Contextual differences

'Shall' is suitable for formal proposals, methodological discussions, and when outlining requirements or objectives.

'Will' usage is more versatile. It is used for casual predictions, descriptions of planned actions, promises, and offers. They are used in a wide range of contexts, from informal discussions to formal predictions about research directions.

Figure 21.5: Forward (Vecteezy Library)

Specific connotations of 'shall'

Obligation or Requirement:
   The solution shall satisfy the boundary conditions. 😊
Strong Intent or Determination. In contexts where the author wishes

---

[1]Contrast this with 'shall'.

to emphasize a strong intention or determination, 'shall' can be more forceful than 'will'.

We shall prove the following theorem within the constraints of the given parameters.

Formality: 'Shall' adds a level of formality to the text.

In this paper, we shall explore the implications of these findings.

21.8   *Foreword vs. forward*

The words 'foreword' and 'forward' sound similar but have entirely different meanings and uses. Understanding and correctly applying these terms can significantly enhance the quality of academic writing, including math texts, books, and papers.

Foreword

A 'foreword' is a short introductory section in a book, usually written by someone other than the author.

The foreword by a renowned mathematician highlights the unique approach the book takes in explaining complex algebraic structures.

An example of a discussion of a foreword:

The book begins with a foreword by a Nobel Laureate, who shares personal anecdotes about the late 20th-century mathematicians featured in the book and their contributions to the advancement of number theory. This foreword sets the stage for readers, providing insight into the math community's dynamics during this pivotal period.

Forward

'Forward' is an adjective, adverb, or verb that denotes direction, progress, or movement towards the front or ahead in time.

Moving forward, we will consider the implications of this theorem for modern cryptography.

———

In moving forward with our analysis, we examine the impact of recent findings in computational complexity on traditional number theory, specifically focusing on how these advancements challenge our understanding of prime numbers.

As we look forward to next week's lecture, we will explore the evolution of algebraic geometry and its role in solving equations that have puzzled mathematicians for centuries. 😊

The team decided to move forward with the proposed algorithm, testing it across various datasets to evaluate its efficiency in solving nonlinear optimization problems. 😊

21.9   Not vs. rather than

The distinction between 'not' and 'rather than' in English is crucial.

Not

'Not' is a negation word used to make a sentence, clause, or phrase negative.

The sequence is not convergent. 😊
'Not' negates "convergent".

In this lecture, we will explore functions that are not differentiable at certain points, such as the absolute value function at zero. 😊
'Not' negates "differentiable at certain points".

The book was filled with counterexamples.

The book was not filled with counterexamples. 😊

Figure 21.6: Book of examples (Vecteezy Library)

Rather than

'Rather than' is a comparative phrase used to express a preference for one option over another. It is often used to contrast two choices, highlighting a preferred one.

Use the method of math induction rather than a direct proof for demonstrating the theorem. 😊

In analyzing the topological structure of the data, the study employs a nonlinear dimensionality reduction technique rather than a traditional linear method. 😊

―――――

During the seminar, the speaker argued for focusing on probabilistic proofs rather than deterministic ones when dealing with algorithms in computational complexity. 😊

Rather than vs. instead of

Examples:

He chose to walk rather than drive. 😊

Let's go to the park instead of the mall. 😊

When followed by a verb, 'instead of' uses the -ing form of the verb.

She went to the gym instead of staying at home. 😊

When followed by a verb, 'rather than' uses the -ing form of the verb.

She went to the gym instead of staying at home. 😊

When followed by a verb, 'instead of' uses the base form of the verb.

She decided to take the bus rather than drive to work. 😊

Figure 21.7: During the seminar (Vecteezy Library)

21.10   Cannot vs. can not

The distinction between 'cannot' and 'can not' in English is subtle yet significant.

Cannot

'Cannot' is a contraction that denotes impossibility. It is the most commonly used form to express that something is not possible or permissible.

A prime number cannot be divided evenly by any number except 1 and itself.

Can not

'Can not' is less commonly used and may be seen as a variant of 'cannot'. However, it can be used to emphasize the 'not' part, especially in contexts where the negation is part of a larger construct.

Specific Contexts: 'Can not' might be used for emphasis or in situations where the action being negated is part of an optional choice.

You can not only solve this equation using numerical methods but also analytically.

You can solve this equation not only using... but also...

Usage

Use 'cannot' for clear statements of impossibility.

'Can not' is rarely needed in math writing but could be used for stylistic reasons or in specific rhetorical constructions to highlight alternatives or choices.

Prefer 'cannot'. For most math writing, 'cannot' is the preferred form.

21.11   Note vs. notice

Distinguishing between 'note' and 'notice'.

Note

'Note' is often used to introduce a point of importance or an observation that should be considered or remembered. It is used to direct the reader's attention to something particular that is crucial for understanding the discussion that follows.

Example:

Note that in the sequence of prime numbers, each number is greater than 1 and divisible only by 1 and itself. 😊

Notice

'Notice', on the other hand, suggests a more passive observation. It invites the reader to observe a pattern which is not in the main focus.

Example:

Upon examining the Fibonacci sequence, one might notice that the ratio of successive terms approaches the golden ratio as the terms increase. 😊

Which one to use

Use 'note' for emphasis and 'notice' for discovery.

21.12   Effect vs. affect

Figure 21.8: Effect and affect (Vecteezy Library)

Both terms, 'effect' and 'affect' relate to change, but they convey different meanings.

Effect

'Effect' is primarily used as a noun (it is also a verb) and refers to the result of a particular action. In math, it often describes the result of a math operation, the impact of a particular theorem, or the outcome of applying a specific algorithm.

The "butterfly effect" in chaos theory illustrates how small changes in initial conditions can have large effects on complex systems. 😊

———

The introduction of a new variable had a significant effect on the outcome of the experiment. 😊

———

The study aims to explore the effect of computational limitations on algorithm efficiency. 😊

This sentence discusses the outcome (effect) of computational limitations on how well algorithms perform.

Sometimes, we use 'to effect' as a verb to mean "to carry out":

It is necessary to carry out all details of the computation. 😊

———

We need to effect a reduction in the amount of homework assigned to students to ensure they have adequate time for other activities.

In this context, 'to effect' means to cause the reduction to happen.

Affect

'Affect' is a verb: to influence or make a difference to something.

It is used in math to describe how one variable influences another or how changes in parameters can alter the outcomes of math models or equations.

Changes in the interest rate affect the overall stability of the financial model. 😊

———

Changing the parameters of the model significantly affects the precision of the results. 😊

———

The choice of numerical methods can greatly affect the convergence rate of iterative solvers. 😊

21.13   *Contain vs. include*

Distinguishing between 'contain' and 'include'.

Contain

The term 'contain' is used to indicate membership. It signifies that an element is in a set. The mathematical symbol for containment is ε.
 Example: The set A contains the element x if $x \varepsilon A$.
 Incorrect: The set A includes the element x.
 Correct: The set A contains the element x.

Include

The term 'include' is used to denote a subset relation. It indicates that all elements of one set are also elements of another set. The symbol for inclusion is $\subseteq$.
 Example: The set A includes the set B if $B \subseteq A$.
 Incorrect: The set A contains the set B.
 Correct: The set A includes the set B.

Examples in Context

Contain: Consider the set of natural numbers N. We say that N contains the number 5 because $5 \in N$.
 Include: Consider two sets, $A = \{1, 2, 3\}$ and $B = \{1, 2\}$. We say that A includes B because $B \subseteq A$.

Practical Tips

Use 'contain' when referring to specific elements within a set. Use 'include' when referring to one set being a subset of another. By adhering to these distinctions, your mathematical writing will be clearer and more precise.

Little modification

To modify

a noun we use 'adjectives';

a verb, an adjective, or another adverb, we use 'adverbs'.

Modifiers play a crucial role in conveying precise meanings in math texts. They specify the properties and relationships of math entities. For example, terms like "continuous", "differentiable", and "integrable" provide essential information about the characteristics of a function. Adverbs can modify verbs or adjectives to fine-tune their meaning, as in "uniformly convergent", where "uniformly" clarifies the nature of "convergence".

Guidelines for using modifiers

Accuracy and necessity Use only necessary modifiers that convey the correct math meaning. Choose precise modifiers. The difference between "increasing" and "strictly increasing" can be crucial.

Supplementary Information The online version contains supplementary material available at https://doi.org/10.1007/978-3-031-65161-8_22.

Placement Place modifiers close to the words they modify to avoid ambiguity. Avoid dangling modifiers that do not clearly refer to any word in the sentence.

Consistency Maintain consistent terminology throughout a document.

Conciseness Opt for the fewest number of words that can clearly convey the intended math idea while maintaining accuracy.

Examples

Use of adjectives

> *The function is very unique in its behavior.* 😟

> *The function exhibits unique behavior.* 🙂

'Unique' is an absolute property, making 'very' redundant.

Adjective misplacement

> *We consider a small complex number z.* 😟

> *We consider a complex number z with small magnitude.* 🙂

Clarifies that "small"' refers to the magnitude of the complex number.

Adverbial clarity

> *The sequence converges nearly uniformly.* 😟

> *The sequence nearly converges uniformly.* 🙂

Specifies that the sequence almost meets the criteria for uniform convergence.

Consistency in modifiers

> *Assume you want to select students for a group project in your class. If we select three students from a class of twenty, how many different groups can we form? Assume the selection is without replacement.* 😟

This statement is incorrect because it talks about selection "independently" but also mentions "without replacement". These are conflicting modifiers because "independently" usually implies each choice is made without any effect from previous choices.

> *Assume you want to select students for a group project in your class. If we select three students from a class of twenty, how many*

different groups can we form? Assume the selection is without replacement. 😕

This corrected statement removes the confusing modifier 'independently'. It clearly states that the selection is without replacement.

Redundant modifiers

The absolutely exact solution is given by... 🙁

The exact solution is given by... 😊

"Exact" already implies precision, making "absolutely" redundant.

Figure 22.1: As x approaches infinity (Vecteezy Library)

Precision in math context

The function grows slowly as x approaches infinity. 🙁

The function grows logarithmically as x approaches infinity. 😊

"Logarithmically" provides a specific description of the growth pattern.

Avoiding ambiguity

The high dimensional vector spaces. 🙁

The high-dimensional vector spaces.

The hyphen clarifies that "high-dimensional" is a compound adjective.

Dangling modifiers

Using the quadratic formula, the equation was solved.

Using the quadratic formula, we solved the equation.

Clarifies who is performing the action, avoiding the implication that the formula itself solved the equation.

Adverb placement

The function strictly monotonically increases.

The function increases strictly monotonically.

Positions the adverb phrase to clearly modify the verb.

Little only

Figure 23.1: Only the Lonely (Vecteezy Library)

"Only the Lonely" ("Know the Way I Feel") is a 1960 song written by Roy Orbison and Joe Melson.

Such a simple word: only.

So much to be careful about!

'Only' is an adverb used to signify that something is limited to the

Supplementary Information The online version contains supplementary material available at https://doi.org/10.1007/978-3-031-65161-8_23.

G. Grätzer, *The Little Book of Math into English*, https://doi.org/10.1007/978-3-031-65161-8_23

specific case mentioned; nothing else applies. The placement of 'only' can define the scope of a theorem, the uniqueness of a solution, or the specific conditions under which a statement holds true.

Only before the verb

When 'only' precedes the verb, it restricts the action of the verb to the condition or object that follows.

John only has 5 *apples. If he gives* 2 *apples to Sarah, how many apples does John have left?* 😊

In this sentence, 'only has' indicates that John possesses no more than 5 apples initially.

Only before the subject

Placing 'only' immediately before the subject or object emphasizes exclusivity.

The only factors of 17 *are* 1 *and prime numbers.* 😊

In this sentence, "Only prime numbers" specifies that the subject, which is "prime numbers", is exclusively the set of numbers that can be considered when discussing factors of 17. Since 17 is a prime number itself, the only factors are 1 and 17, which highlights the exclusivity implied by "only" in this context.

Only before the object

When 'only' precedes the object, it limits the statement exclusively to that object.

The theorem applies to integer values only. 😊

This states that the theorem is not applicable to non-integer values.

Misplacing only

Misplacing 'only' in math contexts can subtly change the meaning of a statement, potentially leading to confusion.

Examples:

Only students who studied passed the exam. 😊

Meaning: Students who studied are the exclusive group that passed the exam.

Misplaced 'only':

Students who only studied passed the exam. 😵

Meaning: The students who passed did nothing but study, implying they had no other activities.

———

Only prime numbers have exactly two distinct positive divisors.

Meaning: Prime numbers are the exclusive set with this property.
Prime numbers only have exactly two distinct positive divisors.☹

Meaning: Prime numbers have no other properties beyond having exactly two distinct positive divisors, which is not accurate.

More illustrations.

Only students in the advanced class can solve this problem in under an hour. ☺

The problem can be solved in under an hour exclusively by students in the advanced class, implying no other students can do so.

Variant:

Students in the advanced class can only solve this problem in under an hour. ☺

When it comes to solving this problem, doing so in under an hour is the only thing students in the advanced class can achieve, implying they cannot solve more complex problems or cannot take longer.

Figure 23.2: The teacher only explained the theorem in the last class (Vecteezy Library)

The teacher only explained the theorem in the last class. 😊

The only action the teacher took during the last class was explaining the theorem, suggesting that no other topics were covered.

Variant:

The teacher explained only the theorem in the last class. 😊

The theorem was the only topic explained by the teacher in the last class, suggesting that other possible topics were not discussed.

Only when the sequence converges to zero can the theorem be applied. 😊

The theorem's applicability is strictly limited to the cases when the sequence converges to zero.

Variant:

When the sequence converges, only zero can be the limit for the theorem to be applied. 😊

Zero is the only value to which the sequence can converge for the theorem to be applicable.

Only half of the students passed the final exam on their first attempt. 😊

Half the students **were not** able to pass the final on their first attempt, suggesting no further attempts were considered.

Little grammar

Not really. This is not 'Grammar 101' but 'Grammatical Terms 101'. See the picture of the big fat grammar books in the Introduction. We do not go there. However, in our discussions, we often have to use some grammatical terms. This appendix explains them.

Figure A.1: Parts of speech (Vecteezy Library)

117
G. Grätzer, *The Little Book of Math into English*, https://doi.org/10.1007/978-3-031-65161-8

A.1   Grammatical terms

Here is a short list.

Noun Names a person, place, thing, or idea.

Equation—Refers to a mathematical statement that asserts the equality between two expressions.

Graph—Represents a diagram showing the relationship between varying quantities, typically using lines or curves.

Pronoun Used in place of a noun.

It—Used to refer to a mathematical object previously mentioned.

Consider a matrix. It is invertible. 😊

They—Can refer to a set of numbers or variables.

If the roots are real, they must be positive. 😊

Verb Indicates action, occurrence, or state of being.

Integrate—The action of finding the integral of a function.

Differentiate—The action of finding the derivative of a function.

Adjective Describes or modifies a noun.

Convergent— Describes a sequence that approaches a specific limit.

Irreducible—Describes a polynomial that cannot be factored into polynomials of lower degree over the same coefficient field.

Adverb Modifies a verb, an adjective, or another adverb, often indicating manner, time, place, or degree.

Quickly—Describes the rate of convergence in a sequence. The sequence converges quickly to zero.

Alternatively—Indicates another method or approach in solving a problem. Alternatively, one might use a geometric approach.

Preposition Shows the relationship between a noun (or pronoun) and other words in a sentence, often indicating position or direction.

Over—Used in expressions like summation over an index.

Within—Indicates a range or a boundary, such as values within the interval.

Conjunction Connects words, phrases, clauses, or sentences (e.g., and, but, or).

And—Links conditions or statements, e.g., $x > 0$ and $x < 10$.

Or—Offers alternatives in conditions, e.g., $x < 0$ or $x > 2$.

Interjection An exclamation or short utterance that expresses emotion but has no grammatical connection to other words.

Oops!—Might be used when an error is realized in a calculation.

Aha!—Exclaimed when a solution or pattern is suddenly understood.

Subject The person, place, thing, or idea that is doing or being something in a sentence. A word or a phrase.

The function—In a statement like "The function is continuous."

This matrix—As in "This matrix is singular."

Predicate Tells something about the subject, usually containing a verb and providing information about the action or state of being of the subject. A word or a phrase.

is prime—Describing the nature of a number.

increases monotonically—Describing the behavior of a function.

Clause A group of words containing a subject and a predicate. It can be independent (main clause) or dependent (subordinate clause).

If x is positive—A condition in a larger mathematical statement.

unless the discriminant is zero—A limiting or conditional modification.

Phrase A group of words that act as a single unit in a sentence but does not have both a subject and a verb.

Greater than zero—A description not forming a full sentence by itself.

Without loss of generality—A common mathematical phrase used to simplify the assumptions for a proof without affecting generality.

Tense Solved—Past tense, used to describe problems that have been solved.

Will determine—Future tense, used for what a particular operation will yield.

Shows the time of the action or state of being indicated by the verb (past, present, future).

Article A word used to modify a noun, which is grammatically necessary for noun phrase structure (definite article: the; indefinite articles: a, an).

the—As in The Integral Test

a—As in A solution to the equation

Syntax The arrangement of words and phrases to create well-formed sentences.

The arrangement in: Let x be a number such that $x > 0$.

The order in: To each pair (x, y), assign the value $x + y$.

Punctuation Symbols that help to structure and organize writing.

Period (.)—Marks the end of a sentence.

Comma (,)—Indicates a pause between parts of a sentence or separates items in a list.

Question Mark (?)—Ends a sentence that asks a question.

Exclamation Mark (!)—Ends a sentence that expresses strong feelings or commands.

Colon (:)—Introduces a list, a quote, or an explanation.

Semicolon (;)—Links independent clauses in a sentence or separates items in a list that already includes commas.

Apostrophe (')—Indicates possession or the omission of letters.

Quotation Marks (" ")—Encloses direct speech, quotations, or titles.

Hyphen (-)—Joins words or parts of words (e.g., twenty-three, well-known).

Dash (—)—Indicates a range or a pause stronger than a comma.

Parentheses (())—Encloses additional information or clarifications.

Brackets [ ]—Used for technical explanations or to clarify meaning.

Ellipsis (...)—Indicates the omission of words or a trailing off of thought.

Sentence types Classifications based on structure (simple, compound, complex, compound-complex) and purpose (declarative, interrogative, imperative, exclamatory).

Declarative: The sum of angles in a triangle is 180 degrees.
Imperative: Calculate the volume of the cylinder.
Interrogative: What is the derivative of the function?
Exclamatory: How elegant this proof is!

A.2 Analyzing sentences

We can practice the parts of speech best by analyzing sentences. First, we analyze:

The function is continuous and differentiable over the interval.

The subject is "The function" and the predicate is "is continuous and differentiable over the interval".

The—Definite article. It specifies a particular instance of the noun that follows, clarifying that it is a specific function.

function—Noun. The subject of the sentence. It is the thing being described, in this case, a math function.

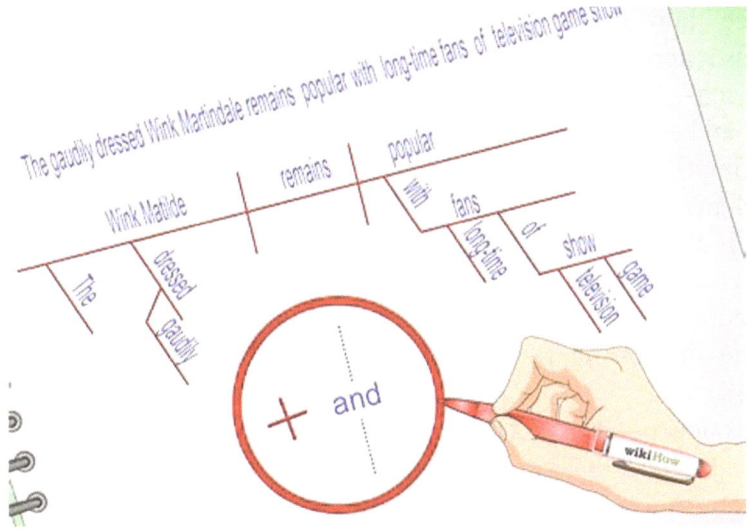

Figure A.2: Analyzing a sentence (Vecteezy Library)

is—Verb (linking). This linking verb connects the subject (function) to the adjectives (continuous and differentiable) that describe it. The verb 'is' is a form of the verb 'to be', used here to denote existence or a state of being.

continuous—Adjective. Describes a property of the function.

and—Conjunction. Joins the adjectives 'continuous' and 'differentiable'. This indicates that both characteristics (continuous and differentiable) apply to the function.

differentiable—Adjective. Another descriptive term for the function, indicating that it can be differentiated.

over—Preposition. Introduces the prepositional phrase "over the interval", indicating the domain or range within which the function is continuous and differentiable.

the—Definite article. Used before 'interval' to specify which interval is being discussed. It points to a particular interval known to the speaker and the listener or reader.

interval—Noun. The object of the preposition 'over'. It specifies the range of values for which the function's properties of being continuous and differentiable apply.

Second, we analyze the sentence:

Each element of the sequence converges to zero as n approaches infinity.

The subject is "Each element of the sequence" and the predicate is "converges to zero as n approaches infinity!".

Each—Pronoun. Specifies every individual member of a group or set, in this case, referring to elements of a sequence.

Figure A.3: Analyzing a second sentence (Vecteezy Library)

element—Noun. The subject of the sentence. It's what the sentence is about, referring to a member of the math sequence.

of—Preposition. Shows the relationship between 'element' and 'the sequence', indicating that the elements belong to the sequence.

the—Determiner. Specifies a particular sequence that is being discussed.

sequence—Noun. The object of the preposition 'of'. It specifies the math object that contains the elements.

converges—Verb. Describes the action of the elements of the sequence. It means that the elements approach a specific value, in this context, zero.

to—Preposition. Indicates direction or movement towards something, in this case, zero.

zero—Noun. The preposition 'to' acts on it; it indicaties the value to which the sequence's elements are converging.

as—Conjunction. Introduces a subordinate clause 'n approaches infinity', specifying the condition under which the convergence occurs.

n—Noun. Subject of the subordinate clause.

approaches—Verb. Describes the action of the variable 'n'.

infinity—Noun. The object of the verb 'approaches', specifying the limit that the variable 'n' moves towards.

Such an analysis is often illustrated by graphs, called Dependency Parse Trees, Constituency Parse Trees, Semantic Networks, Co-occurrence Networks, and Transition Networks.

Here are two examples of diagrams.

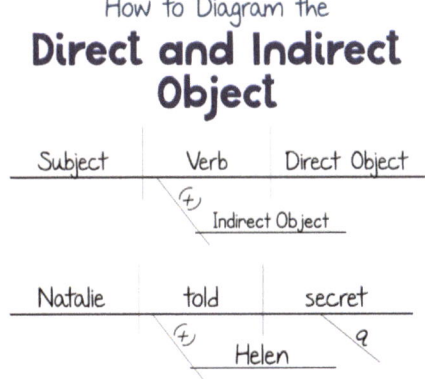

Figure A.4: A direct and indirect object diagram (Vecteezy Library)

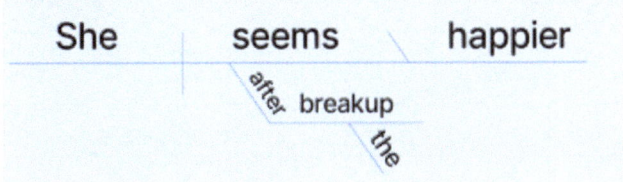

Figure A.5: A sentence diagram (Vecteezy Library)

B

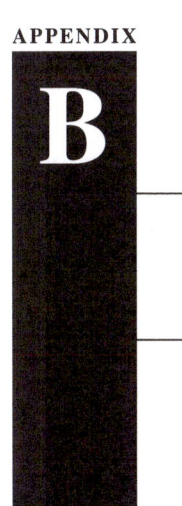

Little ChatGPT for math

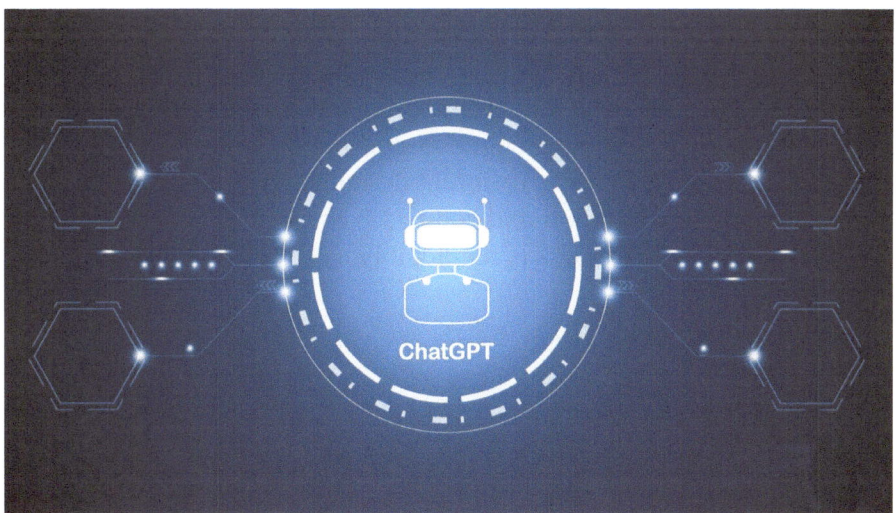

Figure B.1: ChatGPT (Vecteezy Library)

B.1   Introduction

Working on this book, I had an indispensable assistant: ChatGPT Plus, the subscription service of ChatGPT, introduced in the next section. Often, I turned to it for help with English grammar.

In my discussions and email exchanges with experts, skepticism was a common response: "You know my opinion of ChatGPT and other large language models; they can be misleading, and under certain circumstances, produce entirely captivating yet nonsensical and fabricated content."

This response puzzled me. Each time I consulted ChatGPT, it provided clear and unambiguous answers. To help you gain the same benefits, I decided to write a ChatGPT primer for mathematicians.

B.2   What is ChatGPT?

ChatGPT is an LLM (Large Language Model), an artificial intelligence model trained to understand and generate human language. An LLM is 'large' because it is trained on extensive text data, including books, articles, websites, and other digital texts, to grasp language patterns, vocabulary, and grammar.

An LLM can perform various language-related tasks, such as answering questions, completing sentences, and summarizing texts. The OpenAI GPT series (Generative Pre-trained Transformer) are notable examples of LLMs. I have experience only with ChatGPT Plus. Of course, other LLMs would also provide similar assistance.

These large language models have garnered considerable attention for their ability to produce coherent and contextually relevant text, proving invaluable in many different domains ranging from creative writing to technical support and coding.

To start using ChatGPT is remarkably easy. Go to

`openai.com`

and start asking questions.

B.3   What it can't do

It should now be clear that we should ask ChatGPT questions that require the large amount of data it acquired to answer. Otherwise you may get 'nonsensical and fabricated content'.

Donald E. Knuth[1]

`https://www-cs-faculty.stanford.edu/~knuth/chatGPT20.txt`

illustrates it:

"Since one of today's popular recreations is to play with ChatGPT, I decided on 07 April 2023 to try my own little experiment." Knuth writes.

Knuth's first question:

1. Tell me what Donald Knuth says to Stephen Wolfram about ChatGPT.

Clearly, the answer you get does not make sense. There is no large relevant data set. You are asking the shepherd in New Zealand about elephants.

His tenth question is better:

10. How many parts are in The Haj by Leon Uris?

The answer is incorrect (it is divided into three parts, ChatGPT claims). This is interesting. The Haj does not have a pdf version, so it is not part of the data set ChatGPT was trained on.

I asked ChatGPT:

Be an expert on 20th century novels. How many parts are in The Haj by Leon Uris?

And the answer is:

As of my last update in April 2023, I don't have the specific titles of the parts in 'The Haj' by Leon Uris. For the most accurate and detailed information, including the titles of the parts, I recommend referring to a physical copy.

I did:

---

[1]The creator of TEX.

B.4   *Asking good questions*

There are several key considerations for asking effective questions (if you want to charge a lot, call this 'prompt engineering').

1. Specify the field of inquiry. For instance, as demonstrated in Section B.3, we initiated the question with

   Be an expert on 20th century novels.

   If our focus is solely on English grammar, you should begin with

   Be an expert on English grammar.

   Once you instruct ChatGPT in this way, it will remember this command.

2. Asking ChatGPT a question is just the beginning of a conversation. For example, ask:

   Define independent clauses

   and after the long answer, respond

   give math examples

You are reading Topic 17 and want to know whether there is a comma before 'then' in

If the lattice L is modular, then it does not contain M_3 as a sublattice.

It is all explained in Section 17.1, but you do not want to spend the time reading it. So ask ChatGPT:

In the following sentence, do I need a comma before 'it': If the lattice L is modular, then it does not contain M_3 as a sublattice.

ChatGPT answers:

In your sentence, you do not need a comma before 'it' because the sentence structure doesn't require one in this context. r instance), and so on.

B.5   *The GPT Store*

In the second week of January, 2024 (as I am working on this book), Open AI unveiled the GPT Store, see

    https://chat.openai.com/gpts

Reuters reported: "It is OpenAI's attempt to build on the consumer success of ChatGPT, which introduced the world to generative AI last year, dazzling users with its ability to write humanlike prose and

Figure B.2: The GPT Store (Vecteezy Library)

poetry." We call the items in the store Special Purpose GPTs (GPT SPs).

The store also has top picks in seven categories; we discuss only Writing.

B.6   Style and grammar

In the Writing category, you will find 12 items, most of which help you write better. For my work, I use the Creative Writing Coach. I highly recommend it.

B.6.1   A short dialogue for spellchecking

I ask ChatGPT:

Correct it for grammar and style: The courious may wonder, why is the the introduction so short. Are you hiding something their?

Note that there are three errors.

1. "courious" is a spelling error. All spellcheckers should get this.

2. "the the" should be "the". Some spellcheckers would get this.

3. "their" should be "there". No spellcheckers would find this, because the misspelled word "their" is a meaningful word (see the comment on page 83), but ChatGPT does.

ChatGPT answers:

The curious may wonder, why is the introduction so short? Are you hiding something there?

Changes made:

1. *Corrected "courious" to "curious" for spelling accuracy.*

2. *Removed redundant "the" before "introduction".*

3. *Corrected "their" to "there" for the appropriate usage.*

ChatGPT the super proofreader!

B.6.2   A longer editing session

Here is a typical editing dialogue. We ask for help to rewrite an earler version of the Introduction to this book (observe that the prompt asks 'list changes by sentence'; this is very important):

Edit for grammar and style, list changes by sentence:

A typical English grammar book ranges from 400 to 1860 pages in length (the runner up is 1779).

These are general books not particularly useful for math. I think few mathematicians would want to spend time reading any of them. So this book is not trying to teach grammar, but it does utilize some terminology.

What we aim to do is simple and modest: select a few easy topics for discussion. If you follow our recommendations, you will eliminate 80% of your errors, making your papers easier to read and understand.

Note: 80%! Achieving 100% would require 600 pages!

I suggest reading a topic a day. It's just a few minutes of easy reading.

Have fun.

Now the reply. It starts with a general appraisal of my prose.

Your writing has a conversational and engaging tone, which is excellent for drawing in readers. You've effectively communicated the purpose of the book and set a relaxed, approachable tone. Here are some suggestions to enhance the clarity and flow:

Then a sentence by sentence analysis:

Introduction: The opening sentence about English grammar books seems a bit out of context. Consider rephrasing to directly link it to the subject of your book. For example: "Unlike typical English grammar books, which can be extensive and overwhelming, this book adopts a more focused approach."

Purpose and Audience: Clarify early on that the book is for mathematicians, to immediately establish relevance for your target audience.

Encouraging Tone: The part about eliminating 80% of errors is great for motivation. Consider adding a brief explanation of why 100% perfection isn't necessary, to further ease the pressure on the reader.

Call to Action: The suggestion to read a topic a day is excellent. You might want to add a line about the benefits of consistent, daily learning.

Closing: The closing lines are friendly and encouraging. You might want to add a final sentence that ties back to the goal of the book.

And finally, the revised version:

Stepping away from the daunting lengths of typical English grammar books, some spanning from 400 to 1860 pages, this guide presents a concise and mathematician-friendly approach. Our goal isn't to overwhelm you with extensive grammar rules, but to offer you key insights that are directly relevant and easily applicable.

Designed specifically for mathematicians, this book bypasses the complex jargon of standard grammar guides. We aim to simplify the process: selecting a handful of essential topics, ranging from [specific example] to [specific example]. By following our straightforward recommendations, you can reduce up to 80% of common errors in your papers, enhancing their clarity and readability. And don't worry about achieving 100% perfection – that would require an entire 600-page manual!

We suggest a manageable strategy: one topic a day. It's just a few minutes of easy reading that can significantly improve your writing skills.

Have fun and enjoy this journey towards clearer, more effective math writing.

Much improved! This is really ChatGPT on steroids.

B.7   Tokens, darn it

Life would be easier without tokens. But we need them: ChatGPT uses them to measure the size of your prompts and answers, and restricts them to 4096 tokens for ChatGPT 3.5 and twice as many for ChatGPT 4.0.

Some examples of tokens include single words like 'apple', 'beautiful'; 'unbelievable' is tokenized into 'un,' 'believ', and 'able'; sequences of digits are tokenized as single units, such as '2023' or '42'.

Note that 4096 tokens are roughly equivalent to 6 to 8 pages of text.

Figure B.3: Tokens (Vecteezy Library)

To live within the token limits, draft clear and concise prompts. Write specific prompts to maximize your ChatGPT token allowance, especially if you seek a more detailed response. Help ChatGPT avoid generating excessively long responses. For example,

What are the top five major tools to solve differential equations? Be concise and use bullet points.

Sometimes, you may need to provide specific details or elaborate on a query, which can use up more of your tokens. For instance, if you need to summarize a long meeting transcript, consider copying only the essential parts of the discussion. Use this prompt:

Summarize the main decisions in this conversation.

B.8   Conclusions

ChatGPT and the GPT SPs deserve thorough study by you, the reader. Amazon's Kindle Store lists hundreds of books, including *ChatGPT Mastery, ChatGPT for Dummies*, and *The ChatGPT Millionaire*. Even 'prompt engineering' has hundreds of listings. Udemy offers thousands of ChatGPT video courses; prompt engineering has hundreds.

I hope I have succeeded in getting you started.

For a different perspective on the subject, see my article [10].

In digital streams,
Wisdom flows through ChatGPT,
Guiding thoughts to light.

Future bright and clear,
Free from illusions' shadow,
Hope's horizon near.

How to write mathematics by P. R. Halmos

This is a subjective essay, and its title is misleading; a more honest title might be HOW I WRITE MATHEMATICS. It started with a committee of the American Mathematical Society, on which I served for a brief time, but it quickly became a private project that ran away with me. In an effort to bring it under control I asked a few friends to read it and criticize it. The criticisms were excellent; they were sharp, honest, and constructive; and they were contradictory. "Not enough concrete examples," said one; "don't agree that more concrete examples are needed," said another. "Too long" said one; "maybe more is needed" said

Figure C.1: P. R. Halmos

G. Grätzer, *The Little Book of Math into English*, https://doi.org/10.1007/978-3-031-65161-8

another. "There are traditional (and effective) methods of minimizing the tediousness of long proofs, such as breaking them up in a series of lemmas" said one. "One of the things that irritates me greatly is the custom (especially of beginners) to present a proof as a long series of elaborately stated, utterly boring lemmas" said another.

There was one thing that most of my advisors agreed on; the writing of such an essay is bound to be a thankless task. Advisor 1: "By the time a mathematician has written his second paper, he is convinced he knows how to write papers, and would react to advice with impatience." Advisor 2: "All of us, I think, feel secretly that if we but bothered we could be really first rate expositors. People who are quite modest about their mathematics will get their dander up if their ability to write well is questioned." Advisor 3 used the strongest language; he warned me that since I cannot possibly display great intellectual depth in a discussion of matters of technique, I should not be surprised at "the scorn you may reap from some of our more supercilious colleagues".

My advisors are established and well-known mathematicians. A credit line from me here wouldn't add a thing to their stature, but my possible misunderstanding, misplacing, and misapplying their advice might cause them annoyance and embarrassment. That is why I decided on the unscholarly procedure of nameless quotations and the expression of nameless thanks. I am not the less grateful for that, and not the less eager to acknowledge that without their help this essay would have been worse.

"Hier stehe ich; ich kann nicht anders."

C.1   There is no recipe and what it is

I think I can tell someone how to write, but I can't think who would want to listen. The ability to communicate effectively, the power to be intelligible, is congenital, I believe, or, in any event, it is so early acquired that by the time someone reads my wisdom on the subject he is likely to be invariant under it. To understand a syllogism is not something you can learn; you are either born with the ability or you are not. In the same way, effective exposition is not a teachable art; some can do it and some cannot. There is no usable recipe for good writing.

Then why go on? A small reason is the hope that what I said isn't quite right; and, any way, I'd like a chance to try to do what perhaps cannot be done. A more practical reason is that in the other arts that require innate talent, even the gifted ones who are born with it are not usually born with full knowledge of all the tricks of the trade. A few essays such as this may serve to "remind" (in the sense of Plato) the ones who want to be and are destined to be the expositors of the future of the techniques found useful by the expositors of the past.

The basic problem in writing mathematics is the same as in writing biology, writing a novel, or writing directions for assembling a harpsichord: the problem is to communicate an idea. To do so, and to do it clearly, you must have something to say, and you must have someone to say it to, you must organize what you want to say, and you must arrange it in the order you want it said in, you must write it, rewrite it, and re-rewrite it several times, and you must be willing to think hard about and work hard on mechanical details such as diction, notation, and punctuation. That's all there is to it.

C.2 Say something

It might seem unnecessary to insist that in order to say something well you must have something to say, but it's no joke. Much bad writing, mathematical and otherwise, is caused by a violation of that first principle. Just as there are two ways for a sequence not to have a limit (no cluster points or too many), there are two ways for a piece of writing not to have a subject (no ideas or too many). The first disease is the harder one to catch. It is hard to write many words about nothing, especially in mathematics, but it can be done, and the result is bound to be hard to read. There is a classic crank book by Carl Theodore Heisel [13] that serves as an example. It is full of correctly spelled words strung together in grammatical sentences, but after three decades of looking at it every now and then I still cannot read two consecutive pages and make a one-paragraph abstract of what they say; the reason is, I think, that they don't say anything.

The second disease is very common: there are many books that violate the principle of having something to say by trying to say too many things. Teachers of elementary mathematics in the U.S.A. frequently complain that all calculus books are bad. That is a case in point. Calculus books are bad because there is no such subject as calculus; it is not a subject because it is many subjects. What we call calculus nowadays is the union of a dab of logic and set theory, some axiomatic theory of complete ordered fields, analytic geometry and topology, the latter in both the "general" sense (limits and continuous functions) and the algebraic sense (orientation), real-variable theory properly so called (differentiation), the combinatoric symbol manipulation called formal integration, the first steps of low dimensional measure theory, some differential geometry, the first steps of the classical analysis of the trigonometric, exponential, and logarithmic functions, and, depending on the space available and the personal inclinations of the author, some cook-book differential equations, elementary mechanics, and a small assortment of applied mathematics. Any one of these is hard to write a good book on; the mixture is impossible.

Nelson's little gem of a proof that a bounded harmonic function is a constant [16] and Dunford and Schwartz's monumental treatise on functional analysis [7] are examples of mathematical writings that have something to say. Nelson's work is not quite half a page and *Dunford–Schwartz* is more than four thousand times as long, but it is plain in each case that the authors had an unambiguous idea of what they wanted to say. The subject is clearly delineated; it is a subject; it hangs together; it is something to say.

To have something to say is by far the most important ingredient of good exposition—so much so that if the idea is important enough, the work has a chance to be immortal even if it is confusingly misorganized and awkwardly expressed. Birkhoff's proof of the ergodic theorem [2] is almost maximally confusing, and Vanzetti's "last letter" [19] is halting and awkward, but surely anyone who reads them is glad that they were written. To get by on the first principle alone is, however, only rarely possible and never desirable.

C.3   Speak to someone

The second principle of good writing is to write for someone. When you decide to write something, ask yourself who it is that you want to reach. Are you writing a diary note to be read by yourself only, a letter to a friend, a research announcement for specialists, or a textbook for undergraduates? The problems are much the same in any case; what varies is the amount of motivation you need to put in, the extent of informality you may allow yourself, the fussiness of the detail that is necessary, and the number of times things have to be repeated. All writing is influenced by the audience, but, given the audience, an author's problem is to communicate with it as best he can.

Publishers know that 25 years is a respectable old age for most mathematical books; for research papers five years (at a guess) is the average age of obsolescence. (Of course there can be 50-year old papers that remain alive and books that die in five.) Mathematical writing is ephemeral, to be sure, but if you want to reach your audience now, you must write as if for the ages.

I like to specify my audience not only in some vague, large sense (e.g., professional topologists, or second-year graduate students), but also in a very specific, personal sense. It helps me to think of a person, perhaps someone I discussed the subject with two years ago, or perhaps a deliberately obtuse, friendly colleague, and then to keep him in mind as I write. In this essay, for instance, I am hoping to reach mathematics students who are near the beginning of their thesis work, but, at the same time, I am keeping my mental eye on a colleague whose ways can stand mending. Of course I hope that (a) he'll be converted to my ways, but (b) he won't take offence if and when he realizes that I am writing for him.

There are advantages and disadvantages to addressing a very sharply specified audience. A great advantage is that it makes easier the mind reading that is necessary; a disadvantage is that it becomes tempting to indulge in snide polemic comments and heavy-handed "in" jokes. It is surely obvious what I mean by the disadvantage, and it is obviously bad; avoid it. The advantage deserves further emphasis.

The writer must anticipate and avoid the reader's difficulties. As he writes, he must keep trying to imagine what in the words being written may tend to mislead the reader, and what will set him right. I'll give examples of one or two things of this kind later; for now I emphasize that keeping a specific reader in mind is not only helpful in this aspect of the writer's work, it is essential.

Perhaps it needn't be said, but it won't hurt to say, that the audience actually reached may differ greatly from the intended one. There is nothing that guarantees that a writer's aim is always perfect. I still say it's better to have a definite aim and hit something else, than to have an aim that is too inclusive or too vaguely specified and have no chance of hitting anything. Get ready, aim, and aim, and hope that you'll hit a target: the target you were aiming at, for choice, but some target in preference to none.

C.4   *Organize first*

The main contribution that an expository writer can make is to organize and arrange the material so as to minimize the resistance and maximize the insight of the reader and keep him on the track with no unintended distractions. What, after all, are the advantages of a book over a stack of reprints? Answer: efficient and pleasant arrangement, emphasis where emphasis is needed, the indication of interconnections, and the description of the examples and counterexamples on which the theory is based; in one word, organization.

The discoverer of an idea, who may of course be the same as its expositor, stumbled on it helter-skelter, inefficiently, almost at random. If there were no way to trim, to consolidate, and to rearrange the discovery, every student would have to recapitulate it, there would be no advantage to be gained from standing "on the shoulders of giants", and there would never be time to learn something new that the previous generation did not know.

Once you know what you want to say, and to whom you want to say it, the next step is to make an outline. In my experience that is usually impossible. The ideal is to make an outline in which every preliminary heuristic discussion, every lemma, every theorem, every corollary, every remark, and every proof are mentioned, and in which all these pieces occur in an order that is both logically correct and psychologically digestible. In the ideal organization there is a place for everything and everything is in its place. The reader's attention is held because he was told early what to expect, and, at the same time and in apparent contradiction, pleasant surprises keep happening that could not have been predicted from the bare bones of the definitions. The parts fit, and they fit snugly. The lemmas are there when they are needed, and the interconnections of the theorems are visible; and the outline tells you where all this belongs.

I make a small distinction, perhaps an unnecessary one, between organization and arrangement. To organize a subject means to decide what the main headings and subheadings are, what goes under each, and what are the connections among them. A diagram of the organization is a graph, very likely a tree, but almost certainly not a chain. There are many ways to organize most subjects, and usually there are many ways to arrange the results of each method of organization in a linear order. The organization is more important than the arrangement, but the latter frequently has psychological value.

One of the most appreciated compliments I paid an author came from a fiasco; I botched a course of lectures based on his book. The way it started was that there was a section of the book that I didn't like, and I skipped it. Three sections later I needed a small fragment from the end of the omitted section, but it was easy to give a different proof. The same sort of thing happened a couple of times more, but each time a little ingenuity and an ad hoc concept or two patched the leak. In the next chapter, however, something else arose in which what was needed was not a part of the omitted section but the fact that the results of that section were applicable to two apparently very different situations. That was almost impossible to patch up, and after that chaos rapidly set in. The organization of the book was tight; things were there because they were needed; the presentation had the kind of coherence which makes for ease in reading and understanding it. At the same time the wires that were holding it all together were not obtrusive; they became visible only when a part of the structure was tampered with.

Even the least organized authors make a coarse and perhaps unwritten outline; the subject itself is, after all, a one-concept outline of the book. If you know that you are writing about measure theory, then you have a two-word outline, and that's something. A tentative chapter outline is something better. It might go like this: I'll tell them about sets, and then measures, and then functions, and then integrals. At this stage you'll want to make some decisions, which, however, may have to be rescinded later; you may for instance decide to leave probability out, but put Haar measure in.

There is a sense in which the preparation of an outline can take years, or, at the very least, many weeks. For me there is usually a long time between the first joyful moment when I conceive the idea of writing a book and the first painful moment when I sit down and begin to do so. In the interim, while I continue my daily bread and butter work, I daydream about the new project, and, as ideas occur to me about it, I jot them down on loose slips of paper and put them helter-skelter in a folder. An "idea" in this sense may be a field of mathematics I feel should be included, or it may be an item of notation; it may be a proof, it may be an aptly descriptive word, or it may be a witticism that, I hope, will not fall flat but will enliven, emphasize, and exemplify what I want to say. When the painful moment finally arrives, I have the folder at least; playing solitaire with slips of paper can be a big help in preparing the outline.

In the organization of a piece of writing, the question of what to put in is hardly more important than what to leave out; too much detail can be as discouraging as none. The last dotting of the last i, in the manner of the old-fashioned Cours d'Analyse in general and Bourbaki in particular, gives satisfaction to the author who understands it anyway and to the helplessly weak student who never will; for most serious-minded readers it is worse than useless. The heart of mathematics consists of concrete examples and concrete problems. Big general theories are usually afterthoughts based on small but profound insights; the insights themselves come from concrete special cases. The moral is that it's best to organize your work around the central, crucial examples and counterexamples. The observation that a proof proves something a little more general than it was invented for can frequently be left to the reader. Where the reader needs experienced guidance is in the discovery of the things the proof does not prove; what are the appropriate counterexamples and where do we go from here?

C.5 Think about the alphabet

Once you have some kind of plan of organization, an outline, which may not be a fine one but is the best you can do, you are almost ready to start writing. The only other thing I would recommend that you do first is to invest an hour or two of thought in the alphabet; you'll find it saves many headaches later.

The letters that are used to denote the concepts you'll discuss are worthy of thought and careful design. A good, consistent notation can be a tremendous help, and I urge (to the writers of articles too, but especially to the writers of books) that it be designed at the beginning. I make huge tables with many alphabets, with many fonts, for both upper and lower case, and I try to anticipate all the spaces, groups, vectors, functions, points, surfaces, measures, and whatever that will sooner or later need to be baptized. Bad notation can make good exposition bad and bad exposition worse; ad hoc decisions about notation, made mid-sentence in the heat of composition, are almost certain to result in bad notation.

Good notation has a kind of alphabetical harmony and avoids dissonance. Example: either $ax + by$ or $a_1x_1 + a_2x_2$ is preferable to $ax_1 + bx_2$. Or: if you must use Σ for an index set, make sure you don't run into $\sum_{\sigma \in \Sigma} a_\sigma$. Along the same lines: perhaps most readers wouldn't notice that you used $|z| < \epsilon$ at the top of the page and $z \in U$ at the bottom, but that's the sort of near dissonance that causes a vague non-localized feeling of malaise. The remedy is easy and is getting more and more nearly universally accepted: $\in$ is reserved for membership and ϵ for ad hoc use.

Mathematics has access to a potentially infinite alphabet (e.g., x, x' x'', x''', ...), but, in practice, only a small finite fragment of it is usable. One reason is that a human being's ability to distinguish between symbols is very much more limited than his ability to conceive of new ones; another reason is the bad habit of freezing letters. Some old-fashioned analysts would speak of "xyz-space", meaning, I think, 3-dimensional Euclidean space, plus the convention that a point of that space shall always be denoted by "(x, y, z)". This is bad: it "freezes" x, and y, and z, i.e., prohibits their use in another context, and, at the same time, it makes it impossible (or, in any case, inconsistent) to use, say, "(a, b, c)" when "(x, y, z)" has been temporarily exhausted. Modern versions of the custom exist, and are no better. Example: matrices with "property L"—a frozen and unsuggestive designation.

There are other awkward and unhelpful ways to use letters: "CW complexes" and "CCR groups" are examples. A related curiosity that is probably the upper bound of using letters in an unusable way occurs in Lefschetz [15]. There x_i^p is a chain of dimension p (the subscript is just an index), whereas x_p^i is a co-chain of dimension p (and the superscript is an index). Question: what is x_3^2?

As history progresses, more and more symbols get frozen. The standard examples are e, i, and π, and, of course, 0, 1, 2, 3, ... (Who would dare write "Let 6 be a group."?) A few other letters are almost frozen: many readers would feel offended if "n" were used for a complex number, "ε" for a positive integer, and "z" for a topological space. (A mathematician's nightmare is a sequence n_ε that tends to 0 as ε becomes infinite.)

Moral: do not increase the rigid frigidity. Think about the alphabet. It's a nuisance, but it's worth it. To save time and trouble later, think about the alphabet for an hour now; then start writing.

C.6   Write in spirals

The best way to start writing, perhaps the only way, is to write on the spiral plan. According to the spiral plan the chapters get written and re-written in the order 1, 2, 1, 2, 3, 1, 2, 3, 4, etc. You think you know how to write Chapter 1, but after you've done it and gone on to Chapter 2, you'll realize that you could have done a better job on Chapter 2 if you had done Chapter 1 differently. There is no help for it but to go back, do Chapter 1 differently, do a better job on Chapter 2, and then dive into Chapter 3. And, of course, you know what will happen Chapter 3 will show up the weaknesses of Chapters 1 and 2, and there is no help for it ... etc., etc., etc. It's an obvious idea, and frequently an unavoidable one, but it may help a future author to know in advance what he'll run into, and it may help him to know that the same phenomenon will occur not only for chapters, but for sections, for paragraphs, for sentences, and even for words.

The first step in the process of writing, rewriting, and re-rewriting, is writing. Given the subject, the audience, and the outline (and, don't forget, the alphabet), start writing, and let nothing stop you. There is no better incentive for writing a good book than a bad book. Once you have a first draft in hand, spiral-written, based on a subject, aimed at an audience, and backed by as detailed an outline as you could scrape together, then your book is more than half done.

The spiral plan accounts for most of the rewriting and re-rewriting that a book involves (most, but not all). In the first draft of each chapter I recommend that you spill your heart, write quickly, violate all rules, write with hate or with pride, be snide, be confused, be "funny" if you must, be unclear, be ungrammatical—just keep on writing. When you come to rewrite, however, and however often that may be necessary, do not edit but rewrite. It is tempting to use a red pencil to indicate insertions, deletions, and permutations, but in my experience it leads to catastrophic blunders. Against human impatience, and against the all too human partiality everyone feels toward his own words, a red pencil is much too feeble a weapon. You are faced with a first draft that any reader except yourself would find all but unbearable; you must be merciless about changes of all kinds, and, especially, about wholesale omissions. Rewrite means write again every word.

I do not literally mean that, in a 10-chapter book, Chapter 1 should be written ten times, but I do mean something like three or four. The chances are that Chapter 1 should be re-written, literally, as soon as Chapter 2 is finished, and, very likely, at least once again, somewhere after Chapter 4. With luck you'll have to write Chapter 9 only once.

The description of my own practice might indicate the total amount of rewriting that I am talking about. After a spiral-written first draft I usually rewrite the whole book, and then add the mechanical but indispensable reader's aids (such as a list of prerequisites, preface, index, and table of contents). Next, I rewrite again, this time on the typewriter, or, in any event, so neatly and beautifully that a mathematically untrained typist can use this version (the third in some sense) to prepare the "final" typescript with no trouble. The rewriting in this third version is minimal; it is usually confined to changes that affect one word only, or, in the worst case, one sentence. The third version is the first that others see. I ask friends to read it, my wife reads it, my students may read parts of it, and, best of all, an expert junior-grade, respectably paid to do a good job, reads it and is encouraged not to be polite in his criticisms. The changes that become necessary in the third version can, with good luck, be effected with a red pencil; with bad luck they will cause one third of the pages to be retyped. The "final" typescript is based on the edited third version, and, once it exists, it is read, reread, proofread, and reproofread. Approximately two years after it was started (two working years, which may be much more than two calendar years) the book is sent to the publisher. Then begins another kind of labor pain, but that is another story.

Archimedes taught us that a small quantity added to itself often enough becomes a large quantity (or, in proverbial terms, every little bit helps). When it comes to accomplishing the bulk of the world's work, and, in particular, when it comes to writing a book, I believe that the converse of Archimedes' teaching is also true: the only way to write a large book is to keep writing a small bit of it, steadily every day, with no exception, with no holiday. A good technique, to help the steadiness of your rate of production, is to stop each day by priming the pump for the next day. What will you begin with tomorrow? What is the content of the next section to be; what is its title? (I recommend that you find a possible short title for each section, before or after it's written, even if you don't plan to print section titles. The purpose is to test how well the section is planned: if you cannot find a title, the reason may be that the section doesn't have a single unified subject.) Sometimes I write tomorrow's first sentence today; some authors begin today by revising and rewriting the last page or so of yesterday's work. In any case, end each work session on an up-beat; give your subconscious something solid to feed on between sessions. It's surprising how well you can fool yourself that way; the pump-priming technique is enough to overcome the natural human inertia against creative work.

C.7   *Organize always*

Even if your original plan of organization was detailed and good (and especially if it was not), the all-important job of organizing the material does not stop when the writing starts; it goes on all the way through the writing and even after.

The spiral plan of writing goes hand in hand with the spiral plan of organization, a plan that is frequently (perhaps always) applicable to mathematical writing. It goes like this. Begin with whatever you have chosen as your basic concept—vector spaces, say—and do right by it: motivate it, define it, give examples, and give counterexamples. That's Section 1. In Section 2 introduce the first related concept that you propose to study—linear dependence, say—and do right by it: motivate it, define it, give examples, and give counterexamples, and then, this is the important point, review Section 1, as nearly completely as possible, from the point of view of Section 2. For instance: what examples of linearly dependent and independent sets are easily accessible within the very examples of vector spaces that Section 1 introduced? (Here, by the way, is another clear reason why the spiral plan of writing is necessary: you may think, in Section 2, of examples of linearly dependent and independent sets in vector spaces that you forgot to give as examples in Section 1.) In Section 3 introduce your next concept (of course just what that should be needs careful planning, and, more often, a fundamental change of mind that once again makes spiral writing the right procedure), and, after clearing it up in the customary manner, review Sections 1 and 2 from the point of view of the new concept. It works, it works like a charm. It is easy to do, it is fun to do, it is easy to read, and the reader is helped by the firm organizational scaffolding, even if he doesn't bother to examine it and see where the joins come and how they support one another.

The historical novelist's plots and subplots and the detective story writer's hints and dues all have their mathematical analogues. To make the point by way of an example: much of the theory of metric spaces could be developed as a "subplot" in a book on general topology, in unpretentious comments, parenthetical asides, and illustrative exercises. Such an organization would give the reader more firmly founded motivation and more insight than can be obtained by inexorable generality, and with no visible extra effort. As for dues: a single word, first mentioned several chapters earlier than its definition, and then re-mentioned, with more and more detail each time as the official treatment comes closer and closer, can serve as an inconspicuous, subliminal preparation for its full-dress introduction. Such a procedure can greatly help the reader, and, at the same time, make the author's formal work much easier, at the expense, to be sure, of greatly increasing the thought and preparation that goes into his informal prose writing. It's worth it. If you work eight hours to save five minutes of the reader's time, you have saved over 80 man-hours for each 1000 readers, and your name will be deservedly blessed down the corridors of many mathematics buildings. But remember: for an effective use of subplots and dues, something very like the spiral plan of organization is indispensable. The last, least, but still very important aspect of organization that deserves mention here is the correct arrangement of the mathematics from the purely logical point of view. There is not much that one mathematician can teach another about that, except to warn that as the size of the job increases, its complexity increases in frightening proportion. At one stage of writing a 300-page book, I had 1000 sheets of paper, each with a mathematical statement on it, a theorem, a lemma, or even a minor comment, complete with proof. The sheets were numbered, any which way. My job was to indicate on each sheet the numbers of the sheets whose statement must logically come before, and then to arrange the sheets in linear order so that no sheet comes after one on which it's mentioned. That problem had, apparently, uncountably many solutions; the difficulty was to pick one that was as efficient and pleasant as possible.

C.8   *Write good English*

Everything I've said so far has to do with writing in the large, global sense; it is time to turn to the local aspects of the subject.

Why shouldn't an author spell "continuous" as "continous"? There is no chance at all that it will be misunderstood, and it is one letter shorter, so why not? The answer that probably everyone would agree on, even the most libertarian among modem linguists, is that whenever the "reform" is introduced it is bound to cause distraction, and therefore a waste of time, and the "saving" is not worth it. A random example such as this one is probably not convincing; more people would agree that an entire book written in reformed spelling, with, for instance, "izi" for "easy" is not likely to be an effective teaching instrument for mathematics. Whatever the merits of spelling reform may be, words that are misspelled according to currently accepted dictionary standards detract from the good a book can do they delay and distract the reader, and possibly confuse or anger him.

The reason for mentioning spelling is not that it is a common danger or a serious one for most authors, but that it serves to illustrate and emphasize a much more important point. I should like to argue that it is important that mathematical books (and papers, and letters, and lectures) be written in good English style, where good means "correct" according to currently and commonly accepted public standards. (French, Japanese, or Russian authors please substitute "French", "Japanese", or "Russian" for "English".) I do not mean that the style is to be pedantic, or heavy-handed, or formal, or bureaucratic, or flowery, or academic jargon. I do mean that it should be completely unobtrusive, like good background music for a movie, so that the reader may proceed with no conscious or unconscious blocks caused by the instrument of communication and not its content.

Good English style implies correct grammar, correct choice of words, correct punctuation, and, perhaps above all, common sense. There is a difference between "that" and "which", and "less" and "fewer" are not the same, and a good mathematical author must know such things. The reader may not be able to define the difference, but a hundred pages of colloquial misusage, or worse, has a cumulative abrasive effect that the author surely does not want to produce. Fowler [8], Roget [17], and Webster [22] are next to Dunford–Schwartz on my desk; they belong in a similar position on every author's desk. It is unlikely that a single missing comma will convert a correct proof into a wrong one, but consistent mistreatment of such small things has large effects.

The English language can be a beautiful and powerful instrument for interesting, clear, and completely precise information, and I have faith that the same is true for French or Japanese or Russian. It is just as important for an expositor to familiarize himself with that instrument as for a surgeon to know his tools. Euclid can be explained in bad grammar and bad diction, and a vermiform appendix can be removed with a rusty pocket knife, but the victim, even if he is unconscious of the reason for his discomfort, would surely prefer better treatment than that.

All mathematicians, even very young students very near the beginning of their mathematical learning, know that mathematics has a language of its own (in fact it is one), and an author must have thorough mastery of the grammar and vocabulary of that language as well as of the vernacular. There is no Berlitz course for the language of mathematics; apparently the only way to learn it is to live with it for years. What follows is not, it cannot be, a mathematical analogue of Fowler, Roget, and Webster, but it may perhaps serve to indicate a dozen or two of the thousands of items that those analogues would contain.

C.9   Honesty is the best policy

The purpose of using good mathematical language is, of course, to make the understanding of the subject easy for the reader, and perhaps even pleasant. The style should be good not in the sense of flashy brilliance, but good in the sense of perfect unobtrusiveness. The purpose is to smooth the reader's way, to anticipate his difficulties and to forestall them. Clarity is what's wanted, not pedantry; understanding, not fuss.

The emphasis in the preceding paragraph, while perhaps necessary, might seem to point in an undesirable direction, and I hasten to correct a possible misinterpretation. While avoiding pedantry and fuss, I do not want to avoid rigor and precision; I believe that these aims are reconcilable. I do not mean to advise a young author to be ever so slightly but very very cleverly dishonest and to gloss over difficulties. Sometimes, for instance, there may be no better way to get a result than a cumbersome computation. In that case it is the author's duty to carry it out, in public; the best he can do to alleviate it is to extend his sympathy to the reader by some phrase such as "unfortunately the only known proof is the following cumbersome computation".

Here is the sort of thing I mean by less than complete honesty. At a certain point, having proudly proved a proposition p, you feel moved to say: "Note, however, that p does not imply q", and then, thinking that you've done a good expository job, go happily on to other things. Your motives may be perfectly pure, but the reader may feel cheated just the same. If he knew all about the subject, he wouldn't be reading you; for him the non-implication is, quite likely, unsupported. Is it obvious? (Say so.) Will a counterexample be supplied later? (Promise it now.) Is it a standard but for present purposes irrelevant part of the literature? (Give a reference.) Or, horribile dictu, do you merely mean that you have tried to derive q from p, you failed, and you don't in fact know whether p implies q? (Confess immediately!) In any event: take the reader into your confidence.

There is nothing wrong with the often derided "obvious" and "easy to see", but there are certain minimal rules to their use. Surely when you wrote that something was obvious, you thought it was. When, a month, or two months, or six months later, you picked up the manuscript and re-read it, did you still think that that something was obvious? (A few months' ripening always improves manuscripts.) When you explained it to a friend, or to a seminar, was the something at issue accepted as obvious? (Or did someone question it and subside, muttering, when you reassured him? Did your assurance consist of demonstration or intimidation?) The obvious answers to these rhetorical questions are among the rules that should control the use of "obvious". There is another rule, the major one, and everybody knows it, the one whose violation is the most frequent source of mathematical error: make sure that the "obvious" is true.

It should go without saying that you are not setting out to hide facts from the reader; you are writing to uncover them. What I am saying now is that you should not hide the status of your statements and your attitude toward them either. Whenever you tell him something, tell him where it stands: this has been proved, that hasn't, this will be proved, that won't. Emphasize the important and minimize the trivial. There are many good reasons for making obvious statements every now and then; the reason for saying that they are obvious is to put them in proper perspective for the uninitiate. Even if your saying so makes an occasional reader angry at you, a good purpose is served by your telling him how you view the matter. But, of course, you must obey the rules. Don't let the reader down; he wants to believe in you. Pretentiousness, bluff, and concealment may not get caught out immediately, but most readers will soon sense that there is something wrong, and they will blame neither the facts nor themselves, but, quite properly, the author. Complete honesty makes for greatest clarity.

C.10   Down with the irrelevant and the trivial

Sometimes a proposition can be so obvious that it needn't even be called obvious and still the sentence that announces it is bad exposition, bad because it makes for confusion, misdirection, delay. I mean something like this: "If R is a commutative semisimple ring with unit and if x and y are in R, then $x^2 - y^2 = (x - y)(x + y)$." The alert reader will ask himself what semisimplicity and a unit have to do with what he had always thought was obvious. Irrelevant assumptions wantonly dragged in, incorrect emphasis, or even just the absence of correct emphasis can wreak havoc.

Just as distracting as an irrelevant assumption and the cause of just as much wasted time is an author's failure to gain the reader's confidence by explicitly mentioning trivial cases and excluding them if need be. Every complex number is the product of a non-negative number and a number of modulus 1. That is true, but the reader will feel cheated and insecure if soon after first being told that fact (or being reminded of it on some other occasion, perhaps preparatory to a generalization being sprung on him) he is not told that there is something fishy about 0 (the trivial case). The point is not that failure to treat the trivial cases separately may sometimes be a mathematical error; I am not just saying "do not make mistakes". The point is that insistence on legalistically correct but insufficiently explicit explanations ("The statement is correct as it stands what else do you want?") is misleading, bad exposition, bad psychology. It may also be almost bad mathematics. If, for instance, the author is preparing to discuss the theorem that, under suitable hypotheses, every linear transformation is the product of a dilatation and a rotation, then his ignoring of 0 in the 1-dimensional case leads to the reader's misunderstanding of the behavior of singular linear transformations in the general case.

This may be the right place to say a few words about the statements of theorems: there, more than anywhere else, irrelevancies must be avoided.

The first question is where the theorem should be stated, and my answer is: first. Don't ramble on in a leisurely way, not telling the reader where you are going, and then suddenly announce "Thus we have proved that ...". The reader can pay closer attention to the proof if he knows what you are proving, and he can see better where the hypotheses are used if he knows in advance what they are. (The rambling approach frequently leads to the "hanging" theorem, which I think is ugly. I mean something like: "Thus we have proved

THEOREM 2 ...".

The indentation, which is after all a sort of invisible punctuation mark, makes a jarring separation in the sentence, and, after the reader has collected his wits and caught on to the trick that was played on him, it makes an undesirable separation between the statement of the theorem and its official label.)

This is not to say that the theorem is to appear with no introductory comments, preliminary definitions, and helpful motivations. All that comes first; the statement comes next; and the proof comes last. The statement of the theorem should consist of one sentence whenever possible: a simple implication, or, assuming that some universal hypotheses were stated before and are still in force, a simple declaration. Leave the chit-chat out: "Without loss of generality we may assume ... " and "Moreover it follows from Theorem 1 that ... " do not belong in the statement of a theorem.

Ideally the statement of a theorem is not only one sentence, but a short one at that. Theorems whose statement fills almost a whole page (or more!) are hard to absorb, harder than they should be; they indicate that the author did not think the material through and did not organize it as he should have done. A list of eight hypotheses (even if carefully so labelled) and a list of six conclusions do not a theorem make; they are a badly expounded theory. Are all the hypotheses needed for each conclusion? If the answer is no, the badness of the statement is evident; if the answer is yes, then the hypotheses probably describe a general concept that deserves to be isolated, named, and studied.

C.11   Do and do not repeat

One important rule of good mathematical style calls for repetition and another calls for its avoidance. By repetition in the first sense I do not mean the saying of the same thing several times in different words. What I do mean, in the exposition of a precise subject such as mathematics, is the word-for-word repetition of a phrase, or even many phrases, with the purpose of emphasizing a slight change in a neighboring phrase. If you have defined something, or stated something, or proved something in Chapter 1, and if in Chapter 2 you want to treat a parallel theory or a more general one, it is a big help to the reader if you use the same words in the same order for as long as possible, and then, with a proper roll of drums, emphasize the difference. The roll of drums is important. It is not enough to list six adjectives in one definition, and re-list five of them, with a diminished sixth, in the second. That's the thing to do, but what helps is to say, in addition: "Note that the first five conditions in the definitions of p and q are the same; what makes them different is the weakening of the sixth."

Often in order to be able to make such an emphasis in Chapter 2 you'll have to go back to Chapter 1 and rewrite what you thought you had already written well enough, but this time so that its parallelism with the relevant part of Chapter 2 is brought out by the repetition device. This is another illustration of why the spiral plan of writing is unavoidable, and it is another aspect of what I call the organization of the material.

The preceding paragraphs describe an important kind of mathematical repetition, the good kind; there are two other kinds, which are bad.

One sense in which repetition is frequently regarded as a device of good teaching is that the oftener you say the same thing, in exactly the same words, or else with slight differences each time, the more likely you are to drive the point home. I disagree. The second time you say something, even the vaguest reader will dimly recall that there was a first time, and he'll wonder if what he is now learning is exactly the same as what he should have learned before, or just similar but different. (If you tell him "I am now saying *exactly* what I first said on p. 3", that helps.) Even the dimmest such wonder is bad. Anything is bad that unnecessarily frightens, irrelevantly amuses, or in any other way distracts. (Unintended double meanings are the woe of many an author's life.) Besides, good organization, and, in particular, the spiral plan of organization discussed before is a substitute for repetition, a substitute that works much better.

Another sense in which repetition is bad is summed up in the short and only partially inaccurate precept: never repeat a proof. If several steps in the proof of Theorem 2 bear a very close resemblance to parts of the proof of Theorem 1, that's a signal that something may be less than completely understood. Other symptoms of the same disease are: "by the same technique (or method, or device, or trick) as in the proof of Theorem 1 . . .", or, brutally, "see the proof of Theorem 1". When that happens the chances are very good that there is a lemma that is worth finding, formulating, and proving, a lemma from which both Theorem 1 and Theorem 2 are more easily and more clearly deduced.

C.12   *The editorial we is not all bad*

One aspect of expository style that frequently bothers beginning authors is the use of the editorial "we", as opposed to the singular "I", or the neutral "one". It is in matters like this that common sense is most important. For what it's worth, I present here my recommendation. Since the best expository style is the least obtrusive one, I tend nowadays to prefer the neutral approach. That does not mean using "one" often, or ever; sentences like "one has thus proved that ..." are awful. It does mean the complete avoidance of first person pronouns in either singular or plural. "Since p, it follows that q." "This implies p." "An application of p to q yields r." Most (all?) mathematical writing is (should be?) factual; simple declarative sentences are the best for communicating facts.

A frequently effective and time-saving device is the use of the imperative. "To find p, multiply q by r." "Given p, put q equal to r." (Two digressions about "given". (1) Do not use it when it means nothing. Example: "For any given p there is a q." (2) Remember that it comes from an active verb and resist the temptation to leave it dangling. Example: Not "Given p, there is a q", but "Given p, find q".) There is nothing wrong with the editorial "we", but if you like it, do not misuse it. Let "we" mean "the author and the reader" (or "the lecturer and the audience"). Thus, it is fine to say "Using Lemma 2 we can generalize Theorem 1", or "Lemma 3 gives us a technique for proving Theorem 4". It is not good to say "Our work on this result was done in 1969" (unless the voice is that of two authors, or more, speaking in unison), and "We thank our wife for her help with the typing" is always bad.

The use of "I", and especially its overuse, sometimes has a repellent effect, as arrogance or ex-cathedra preaching, and, for that reason, I like to avoid it whenever possible. In short notes, obviously in personal historical remarks, and, perhaps, in essays such as this, it has its place.

C.13   *Use words correctly*

The next smallest units of communication, after the whole concept, the major chapters, the paragraphs, and the sentences are the words. The preceding section about pronouns was about words, in a sense, although, in a more legitimate sense, it was about global stylistic policy. What I am now going to say is not just "use words correctly"; that should go without saying. What I do mean to emphasize is the need to think about and use with care the small words of common sense and intuitive logic, and the specifically mathematical words (technical terms) that can have a profound effect on mathematical meaning.

The general rule is to use the words of logic and mathematics correctly. The emphasis, as in the case of sentence-writing, is not encouraging pedantry; I am not suggesting a proliferation of technical terms with hairline distinctions among them. Just the opposite; the emphasis is on craftsmanship so meticulous that it is not only correct, but unobtrusively so.

Here is a sample: "Prove that any complex number is the product of a non-negative number and a number of modulus 1." I have had students who would have offered the following proof: "$-4i$ is a complex number, and it is the product of 4, which is non-negative, and $-i$, which has modulus 1; q.e.d." The point is that in everyday English "any" is an ambiguous word; depending on context it may hint at an existential quantifier ("have you any wool?", "if anyone can do it, he can") or a universal one ("any number can play"). Conclusion: never use "any" in mathematical writing. Replace it by "each" or "every", or recast the whole sentence.

One way to recast the sample sentence of the preceding paragraph is to establish the convention that all "individual variables" range over the set of complex numbers and then write something like

$$\forall x \exists p \exists u [(p = |p|) \wedge (|u| = 1) \wedge (z = pu)].$$

I recommend against it. The symbolism of formal logic is indispensable in the discussion of the logic of mathematics, but used as a means of transmitting ideas from one mortal to another it becomes a cumbersome code. The author had to code his thoughts in it (I deny that anybody thinks in terms of $\exists$, $\forall$, $\wedge$, and the like), and the reader has to decode what the author wrote; both steps are a waste of time and an obstruction to understanding. Symbolic presentation, in the sense of either the modem logician or the classical epsilontist, is something that machines can write and few but machines can read.

So much for "any". Other offenders, charged with lesser crimes, are "where", and "equivalent", and "if ... then ... if ... then". "Where" is usually a sign of a lazy afterthought that should have been thought through before. "If n is sufficiently large, then $|a_n| < \varepsilon$, where ε is a preassigned positive number"; both disease and cure are clear. "Equivalent" for theorems is logical nonsense. (By "theorem" I mean a mathematical truth, something that has been proved. A meaningful statement can be false, but a theorem cannot; "a false theorem" is self-contradictory). What sense does it make to say that the completeness of L^2 is equivalent to the representation theorem for linear functionals on L^2? What is meant is that the proofs of both theorems are moderately hard, but once one of them has been proved, either one, the other can be proved with relatively much less work. The logically precise word "equivalent" is not a good word for *that*. As for "if ... then ... if ... then", that is just a frequent stylistic bobble committed by quick writers and rued by slow readers. "If p, then if q, then r." Logically all is well $(p \Rightarrow (q \Rightarrow r))$, but psychologically it is just another pebble to stumble over, unnecessarily. Usually all that is needed to avoid it is to recast the sentence, but no universally good recasting exists; what is best depends on what is important in the case at hand. It could be "If p and q, then r", or "In the presence of p, the hypothesis q implies the conclusion r", or many other versions.

C.14   Use technical terms correctly

The examples of mathematical diction mentioned so far were really logical matters. To illustrate the possibilities of the unobtrusive use of precise language in the everyday sense of the working mathematician, I briefly mention three examples: function, sequence, and contain.

I belong to the school that believes that functions and their values are sufficiently different that the distinction should be maintained. No fuss is necessary, or at least no visible, public fuss; just refrain from saying things like "the function $z^2 + 1$ is even". It takes a little longer to say "the function f defined by $f(z) = z^2 + 1$ is even", or, what is from many points of view preferable, "the function $z \to z^2 + 1$ is even", but it is a good habit that can sometimes save the reader (and the author) from serious blunder and that always makes for smoother reading.

"Sequence" means "function whose domain is the set of natural numbers". When an author writes "the union of a sequence of measurable sets is measurable" he is guiding the reader's attention to where it doesn't belong. The theorem has nothing to do with the firstness of the first set, the secondness of the second, and so on; the *sequence* is irrelevant. The correct statement is that "the union of a countable set of measurable sets is measurable" (or, if a different emphasis is wanted, "the union of a countably infinite set of measurable sets is measurable"). The theorem that "the limit of a sequence of measurable functions is measurable" is a very different thing; there "sequence" is correctly used. If a reader knows what a sequence is, if he feels the definition in his bones, then the misuse of the word will distract him and slow his reading down, if ever so slightly; if he doesn't really know, then the misuse will seriously postpone his ultimate understanding.

"Contain" and "include" are almost always used as synonyms, often by the same people who carefully coach their students that $\in$ and $\subset$ are not the same thing at all. It is extremely unlikely that the interchangeable use of contain and include will lead to confusion. Still, some years ago I started an experiment, and I am still trying it: I have systematically and always, in spoken word and written, used "contain" for $\in$ and "include" for $\subset$. I don't say that I have proved anything by this, but I can report that (a) it is very easy to get used to, (b) it does no harm whatever, and (c) I don't think that anybody ever noticed it. I suspect, but that is not likely to be provable, that this kind of terminological consistency (with no fuss made about it) might nevertheless contribute to the reader's (and listener's) comfort.

Consistency, by the way, is a major virtue and its opposite is a cardinal sin in exposition. Consistency is important in language, in notation, in references, in typography it is important everywhere, and its absence can cause anything from mild irritation to severe misinformation.

My advice about the use of words can be summed up as follows. (1) Avoid technical terms, and especially the creation of new ones, whenever possible. (2) Think hard about the new ones that you must create; consult Roget; and make them as appropriate as possible. (3) Use the old ones correctly and consistently, but with a minimum of obtrusive pedantry.

C.15   Resist symbols

Everything said about words applies, mutatis mutandis, to the even smaller units of mathematical writing, the mathematical symbols. The best notation is no notation; whenever it is possible to avoid the use of a complicated alphabetic apparatus, avoid it. A good attitude to the preparation of written mathematical exposition is to pretend that it is spoken. Pretend that you are explaining the subject to a friend on a long walk in the woods, with no paper available; fall back on symbolism only when it is really necessary. A corollary to the principle that the less there is of notation the better it is, and in analogy with the principle of omitting irrelevant assumptions, avoid the use of irrelevant symbols. Example: "On a compact space every real-valued continuous function f is bounded." What does the symbol "f" contribute to the clarity of that statement? Another example: "If $0 \leq \lim_n \alpha_n^{1/n} = \rho \leq 1$, then $\lim_n \alpha_n = 0$." What does "ρ" contribute here? The answer is the same in both cases (nothing), but the reasons for the presence of the irrelevant symbols may be different. In the first case "f" may be just a nervous habit; in the second case "ρ" is probably a preparation for the proof. The nervous habit is easy to break. The other is harder, because it involves more work for the author. Without the "ρ" in the statement, the proof will take a half line longer; it will have to begin with something like "Write $\rho = \lim_n \alpha_n^{1/n}$." The repetition (of "$\lim_n \alpha_n^{1/n}$") is worth the trouble; both statement and proof read more easily and more naturally.

A showy way to say "use no superfluous letters" is to say "use no letter only once". What I am referring to here is what logicians would express by saying "leave no variable free". In the example above, the one about continuous functions, "f" was a free variable. The best way to eliminate that particular "f" is to omit it; an occasionally preferable alternative is to convert it from free to bound. Most mathematicians would do that by saying "If f is a real-valued continuous function on a compact space, then f is bounded." Some logicians would insist on pointing out that "f" is still free in the new sentence (twice), and technically they would be right. To make it bound, it would be necessary to insert "for all f" at some grammatically appropriate point, but the customary way mathematicians handle the problem is to refer (tacitly) to the (tacit) convention that every sentence is preceded by all the universal quantifiers that are needed to convert all its variables into bound ones.

The rule of never leaving a free variable in a sentence, like many of the rules I am stating, is sometimes better to break than to obey. The sentence, after all, is an arbitrary unit, and if you want a free "f"

dangling in one sentence so that you may refer to it in a later sentence in, say, the same paragraph, I don't think you should necessarily be drummed out of the regiment. The rule is essentially sound, just the same, and while it may be bent sometimes, it does not deserve to be shattered into smithereens.

There are other symbolic logical hairs that can lead to obfuscation, or, at best, temporary bewilderment, unless they are carefully split. Suppose, for an example, that somewhere you have displayed the relation

$$(*) \qquad \int_0^1 |f(x)|^2 \, dx < \infty,$$

as, say, a theorem proved about some particular f. If, later, you run across another function g with what looks like the same property, you should resist the temptation to say "g also satisfies $(*)$". That's logical and alphabetical nonsense. Say instead "$(*)$ remains satisfied if f is replaced by g", or, better, give $(*)$ a name (in this case it has a customary one) and say "g also belongs to $L^2(0,1)$".

What about "inequality $(*)$", or "equation (7)", or "formula (iii)" ; should all displays be labelled or numbered? My answer is no. Reason: just as you shouldn't mention irrelevant assumptions or name irrelevant concepts, you also shouldn't attach irrelevant labels. Some small part of the reader's attention is attracted to the label, and some small part of his mind will wonder why the label is there. If there is a reason, then the wonder serves a healthy purpose by way of preparation, with no fuss, for a future reference to the same idea; if there is no reason, then the attention and the wonder were wasted.

It's good to be stingy in the use of labels, but parsimony also can be carried to extremes. I do not recommend that you do what Dickson once did [6]. On p. 89 he says: "Then ... we have (1) ..."—but p. 89 is the beginning of a new chapter, and happens to contain no display at all, let alone one bearing the label (1). The display labelled (1) occurs on p. 90, overleaf, and I never thought of looking for it *there*. That trick gave me a helpless and bewildered five minutes. When I finally saw the light, I felt both stupid and cheated, and I have never forgiven Dickson.

One place where cumbersome notation quite often enters is in mathematical induction. Sometimes it is unavoidable. More often, however, I think that indicating the step from 1 to 2 and following it by an airy "and so on" is as rigorously unexceptionable as the detailed computation, and much more understandable and convincing. Similarly, a general statement about $n \times n$ matrices is frequently best proved not by the exhibition of many a_{ij}'s, accompanied by triples of dots laid out in rows and columns and diagonals, but by the proof of a typical (say 3×3) special case.

There is a pattern in all these injunctions about the avoidance of notation. The point is that the rigorous concept of a mathematical proof can be taught to a stupid computing machine in one way only, but to a human being endowed with geometric intuition, with daily increasing experience, and with the impatient inability to concentrate on repetitious detail for very long, that way is a bad way. Another illustration of this is a proof that consists of a chain of expressions separated by equal signs. Such a proof is easy to write. The author starts from the first equation, makes a natural substitution to get the second, collects terms, permutes, inserts and immediately cancels an inspired factor, and by steps such as these proceeds till he gets the last equation. This is, once again, coding, and the reader is forced not only to learn as he goes, but, at the same time, to decode as he goes. The double effort is needless. By spending another ten minutes writing a carefully worded paragraph, the author can save each of his readers half an hour and a lot of confusion. The paragraph should be a recipe for action, to replace the unhelpful code that merely reports the results of the act and leaves the reader to guess how they were obtained. The paragraph would say something like this: "For the proof, first substitute p for q, then collect terms, permute the factors, and, finally, insert and cancel a factor r."

A familiar trick of bad teaching is to begin a proof by saying: "Given ε, let δ be $\left(\frac{\varepsilon}{3M^2+2}\right)^{1/2}$". This is the traditional backward proof-writing of classical analysis. It has the advantage of being easily verifiable by a machine (as opposed to *understandable* by a human being), and it has the dubious advantage that something at the end comes out to be less than ε, instead of less than, say, $\left(\frac{(3M^2+7)\varepsilon}{24}\right)^{1/3}$. The way to make the human reader's task less demanding is obvious: write the proof forward. Start, as the author always starts, by putting something less than ε, and then do what needs to be done—multiply by $3M^2 + 7$ the right time and divide by 24 later, etc., etc.—till you end up with what you end up with. Neither arrangement is elegant, but the forward one is graspable and rememberable.

C.16   Use symbols correctly

There is not much harm that can be done with non-alphabetical symbols, but there too consistency is good and so is the avoidance of individually unnoticed but collectively abrasive abuses. Thus, for instance, it is good to use a symbol so consistently that its verbal translation is always the same. It is good, but it is probably impossible; nonetheless it's a better aim than no aim at all. How are we to read "$\in$": as the verb phrase "is in" or as the preposition "in"? Is it correct to say: "For $x \in A$, we have $x \in B$," or "If $x \in A$, then $x \in B$"? I strongly prefer the latter (always read "$\in$" as "is in") and I doubly deplore the former (both usages occur in the same sentence). It's easy to write and it's easy to read "For x in A, we have $x \in B$"; all dissonance and all even momentary ambiguity is avoided. The same is true for "$\subset$" even though the verbal translation is longer, and even more true for "$\leq$". A sentence such as "Whenever a positive number is ≤ 3, its square is ≤ 9" is ugly.

Not only paragraphs, sentences, words, letters, and mathematical symbols, but even the innocent looking symbols of standard prose can be the source of blemishes and misunderstandings; I refer to punctuation marks. A couple of examples will suffice. First: an equation, or inequality, or inclusion, or any other mathematical clause is, in its informative content, equivalent to a clause in ordinary language, and, therefore, it demands just as much to be separated from its neighbors. In other words: punctuate symbolic sentences just as you would verbal ones. Second: don't overwork a small punctuation mark such as a period or a comma. They are easy for the reader to overlook, and the oversight causes backtracking, confusion, delay. Example: "Assume that $a \in X$. X belongs to the class $C, \dots$". The period between the two X's is overworked, and so is this one: "Assume that X vanishes. X belongs to the class $C, \dots$". A good general rule is: never start a sentence with a symbol. If you insist on starting the sentence with a mention of the thing the symbol denotes, put the appropriate word in apposition, thus: "The set X belongs to the class $C, \dots$".

The overworked period is no worse than the overworked comma. Not "For invertible X, X^* also is invertible", but "For invertible X, the adjoint X^* also is invertible". Similarly, not "Since $p \neq 0, p \in U$", but "Since $p \neq 0$, it follows that $p \in U$". Even the ordinary "If you don't like it, lump it" (or, rather, its mathematical relatives) is harder to digest than the stuffy sounding "If you don't like it, then lump it"; I recommend "then" with "if" in all mathematical contexts. The presence of "then" can never confuse; its absence can.

A final technicality that can serve as an expository aid, and should be mentioned here, is in a sense smaller than even the punctuation marks, it is in a sense so small that it is invisible, and yet, in another sense, it's the most conspicuous aspect of the printed page. What I am talking about is the layout, the architecture, the appearance of the page itself, of all the pages. Experience with writing, or perhaps even with fully conscious and critical reading, should give you a feeling for how what you are now writing will look when it's printed. If it looks like solid prose, it will have a forbidding, sermony aspect; if it looks like computational hash, with a page full of symbols, it will have a frightening, complicated aspect. The golden mean is golden. Break it up, but not too small; use prose, but not too much. Intersperse enough displays to give the eye a chance to help the brain; use symbols, but in the middle of enough prose to keep the mind from drowning in a morass of suffixes.

C.17   All communication is exposition

I said before, and I'd like for emphasis to say again, that the differences among books, articles, lectures, and letters (and whatever other means of communication you can think of) are smaller than the similarities.

When you are writing a research paper, the role of the "slips of paper" out of which a book outline can be constructed might be played by the theorems and the proofs that you have discovered; but the game of solitaire that you have to play with them is the same.

A lecture is a little different. In the beginning a lecture is an expository paper; you plan it and write it the same way. The difference is that you must keep the difficulties of oral presentation in mind. The reader of a book can let his attention wander, and later, when he decides to, he can pick up the thread, with nothing lost except his own time; a member of a lecture audience cannot do that. The reader can try to prove your theorems for himself, and use your exposition as a check on his work; the hearer cannot do that. The reader's attention span is short enough; the hearer's is much shorter. If computations are unavoidable, a reader can be subjected to them; a hearer must never be. Half the art of good writing is the art of omission; in speaking, the art of omission is nine-tenths of the trick. These differences are not large. To be sure, even a good expository paper, read out loud, would make an awful lecture—but not worse than some I have heard.

The appearance of the printed page is replaced, for a lecture, by the appearance of the blackboard, and the author's imagined audience is replaced for the lecturer by live people; these are big differences. As for the blackboard: it provides the opportunity to make something grow and come alive in a way that is not possible with the printed page. (Lecturers who prepare a blackboard, cramming it full before they start speaking, are unwise and unkind to audiences.) As for live people: they provide an immediate feedback that every author dreams about but can never have.

The basic problems of all expository communication are the same; they are the ones I have been describing in this essay. Content, aim and organization, plus the vitally important details of grammar, diction, and notation—they, not showmanship, are the essential ingredients of good lectures, as well as good books.

C.18   Defend your style

Smooth, consistent, effective communication has enemies; they are called editorial assistants or copyreaders.

An *editor* can be a very great help to a writer. Mathematical writers must usually live without this help, because the editor of a mathematical book must be a mathematician, and there are very few mathematical editors. The ideal editor, who must potentially understand every detail of the author's subject, can give the author an inside but nonetheless unbiased view of the work that the author himself cannot have. The ideal editor is the union of the friend, wife, student, and expert junior-grade whose contribution to writing I described earlier. The mathematical editors of book series and journals don't even come near to the ideal. Their editorial work is but a small fraction of their life, whereas to be a good editor is a full-time job. The ideal mathematical editor does not exist; the friend-wife etc. combination is only an almost ideal substitute.

The *editorial assistant* is a full-time worker whose job is to catch your inconsistencies, your grammatical slips, your errors of diction, your misspellings—everything that you can do wrong, short of the mathematical content. The trouble is that the editorial assistant does not regard himself as an extension of the author, and he usually degenerates into a mechanical misapplier of mechanical rules. Let me give some examples.

I once studied certain transformations called "measure-preserving". (Note the hyphen: it plays an important role, by making a single word, an adjective, out of two words.) Some transformations pertinent to that study failed to deserve the name; their failure was indicated, of course, by the prefix "non". After a long sequence of misunderstood instructions, the printed version spoke of a "nonmeasure preserving transformation". That is nonsense, of course, amusing nonsense, but, as such, it is distracting and confusing nonsense.

A mathematician friend reports that in the manuscript of a book of his he wrote something like "p or q holds according as x is negative or positive". The editorial assistant changed that to "p or q holds according as x is positive or negative", on the grounds that it sounds better that way. That could be funny if it weren't sad, and, of course, very very wrong.

A common complaint of anyone who has ever discussed quotation marks with the enemy concerns their relation to other punctuation. There appears to be an international typographical decree according to which a period or a comma immediately to the right of a quotation is "ugly". (As here: the editorial assistant would have changed that to "ugly." if I had let him.) From the point of view of the logical mathematician (and even more the mathematical logician) the decree makes no sense; the comma or period should come where the logic of the situation forces it to come. Thus,

He said: "The comma is ugly."

Here, clearly, the period belongs inside the quote; the two situations are different and no inelastic rule can apply to both.

Moral: there are books on "style" (which frequently means typographical conventions), but their mechanical application by editorial assistants can be harmful. If you want to be an author, you must be prepared to defend your style; go forearmed into the battle.

C.19   Stop

The battle against copyreaders is the author's last task, but it's not the one that most authors regard as the last. The subjectively last step comes just before; it is to finish the book itself—to stop writing. That's hard.

There is always something left undone, always either something more to say, or a better way to say something, or, at the very least, a disturbing vague sense that the perfect addition or improvement is just around the corner, and the dread that its omission would be everlasting cause for regret. Even as I write this, I regret that I did not include a paragraph or two on the relevance of euphony and prosody to mathematical exposition. Or, hold on a minute!, surely I cannot stop without a discourse on the proper naming of concepts (why "commutator" is good and "set of first category" is bad) and the proper way to baptize theorems (why "the closed graph theorem" is good and "the *Cauchy–Buniakowski–Schwarz* theorem" is bad). And what about that sermonette that I haven't been able to phrase satisfactorily about following a model. Choose someone, I was going to say, whose writing can touch you and teach you, and adapt and modify his style to fit your personality and your subject surely I must get that said somehow.

There is no solution to this problem except the obvious one; the only way to stop is to be ruthless about it. You can postpone the agony a bit, and you should do so, by proofreading, by checking the computations, by letting the manuscript ripen, and then by reading the whole thing over in a gulp, but you won't want to stop any more then than before.

When you've written everything you can think of, take a day or two to read over the manuscript quickly and to test it for the obvious major points that would first strike a stranger's eye. Is the mathematics good, is the exposition interesting, is the language clear, is the format pleasant and easy to read? Then proofread and check the computations; that's an obvious piece of advice, and no one needs to be told how to do it. "Ripening" is easy to explain but not always easy to do it means to put the manuscript out of sight and try to forget it for a few months. When you have done all that, and then re-read the whole work from a rested point of view, you have done all you can. Don't wait and hope for one more result, and don't keep on polishing. Even if you do get that result or do remove that sharp corner, you'll only discover another mirage just ahead.

To sum it all up: begin at the beginning, go on till you come to the end, and then, with no further ado, stop.

C.20   The last word

I have come to the end of all the advice on mathematical writing that I can compress into one essay. The recommendations I have been making are based partly on what I do, more on what I regret not having done, and most on what I wish others had done for me. You may criticize what I've said on many grounds, but I ask that a comparison of my present advice with my past action not be one of them. Do, please, as I say, and not as I do, and you'll do better. Then rewrite this essay and tell the next generation how to do better still.

Bibliography

[1] Barbara Beeton, Hyphenation exception log. TUGboat, **39** (2018), 901–912.
For the most up-to-date version, see
`https://mirrors.ctan.org/info/digests`
`/tugboat/hyphenex/tb0hyf.pdf`

[2] G. D. Birkhoff, Proof of the ergodic theorem. Proc. N.A.S., U.S.A. **17** (1931), 656–660.

[3] Lewis Carroll, Alice in Wonderland. Macmillan Publishers, 1865.

[4] Lewis Carroll, Through the Looking Glass. Macmillan Publishers, 1871.

[5] The Chicago Manual of Style, 17th edition. University of Chicago Press, 2017.

[6] L. E. Dickson, Modern Algebraic Theories. Sanborn, Chicago, 1926.

[7] N. Dunford and J. T. Schwartz, Linear Operators. Interscience, New York, 1958.

[8] H. W. Fowler, A Dictionary of Modern English Usage, The Classic First Edition. Oxford, 2010.

[9] G. Grätzer, My publications. Research Gate
`https://www.researchgate.net/publication/`
`371959806_MyPubli`

[10] G. Grätzer, ChatGPT 101 for LaTeX users. EMS Magazine, September 2024 issue.

[11] P. R. Halmos, How to write mathematics. L'Enseignement mathématique, 1970. An extra for this book.

[12] Ernest Hemingway, For Whom the Bell Tolls. Charles Scribner's Sons., 1940.

[13] C. T. Heisel, The Circle Squared beyond Refutation. Heisel, Cleveland, 1934.

[14] R. Huddleston and G. K. Pullum, The Cambridge Grammar of the English Language. Cambridge University Press, 2002.

[15] S. Lefschetz, Algebraic Topology. AMS, New York, 1942.

[16] E. Nelson, A proof of Liouville's theorem. Proc. A.M.S. **12** (1961), 995.

[17] Roget's International Thesaurus. Crowell, New York, 1946.

[18] J. E. Sandys, A History of Classical Scholarship: From the Sixth Century B. C. to the End of the Middle Ages. C. J. Clay and Sons, London, 1903.

[19] J. Thurber and E. Nugent, The Male Animal, Random House, New York, 1940.

[20] J. Trzeciak, Writing Mathematical Papers in English. European Mathematical Society, 2005.

[21] F. Vivaldi, Mathematical Writing. Springer Verlag, 2014.

[22] Webster's New International Dictionary, second edition. Merriam, Springfield, 1951.